History of Mankind's Greatest Disaster

A Walk Through The Chernobyl Nuclear Catastrophe

By Donald B. Grey

Books you may also like:

Quantum Physics Made Easy:

The Introduction Guide For Beginners Who Flunked Maths And Science In Plain Simple English

What In The World Is Quantum Physics?

Do black holes really exist?

Are string theories made of… strings?

What is the Schrödinger's Cat?

Let's face the fact here, you are NOT A SCIENTIST nor a physician, and yet you are curious about those questions that you have been pondering about.

99.99% of the world's mysteries are yet to be discovered and/or solved.

Why not…

It's time for you to rediscover science?

One of the most compelling draws of the sciences for many people is the potential of discovering something that was not known before. Whether someone's doing it for fame, for fortune, or just for the fun of it, discovering something new, leaving your own personal mark for the rest of humanity's time in the universe, is a tempting prospect for many.

How would you feel about naming a star, and for others to know that you named it? That star would be visible in the sky for the rest of your lifetime, and more than likely for your great-great-great-grandchildren's lifetimes. Your discovery would be immortalized above for the life of the star.

Inside this book you will discover:

-String theory and how it came about

-Black holes and quantum gravity

-If Schrödinger's Cat is really a cat?

-Disagreements between Einstein and Bohr

-The double slit experiment

Attention! Quantum Physics is NOT for everyone!

This book is not for people:

-Who doesn't want to impress their girl with science

-Who are not curious about the universe

-Who isn't inspired to name their own science theory

Read now on Amazon

Billion Dollar Façade:

The Rise And Fall Of Theranos And Elizabeth Holmes

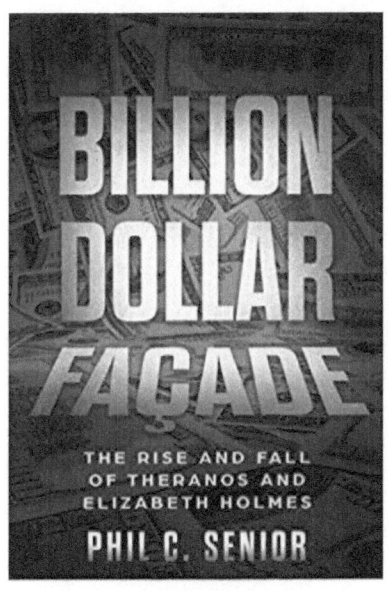

In the fall of 2015, The Wall Street Journal ran an article on its front page which effectively set the biotechnology world and the insular Silicon Valley scene on fire. Theranos, a much hailed startup which was disrupting the world of healthcare and led by a visionary young woman, was a fraud, the story claimed.

To most people, the article amounted to nothing short of blasphemy. To stain the reputation of Elizabeth Holmes and her company, valued at $10 billion, was to align oneself against progress and it was initially seen as the backward thinking that was prevalent in mainstream, non-Silicon Valley society. Here was the next Steve Jobs, right down to the ridiculous black turtleneck, being tarnished much in the manner Galileo was sabotaged by the Catholic Church for simply telling the truth.

The backlash from Theranos had followed similar lines of argument. The story was completely made up and the firm had rebuttals on hand which would disprove every single line of the story. The reporter who broke the story, John Carreyrou, had his name dragged through the mud and the journal was buried with threats of lawsuits from the powerful law firm representing Theranos.However, despite the threats and vows to set things right, there was simply too much smoke around Theranos. It seemed as if the Valley had finally removed its blinders and began taking a look at facts. The facts were damning: Theranos and Elizabeth Holmes were frauds. As of current writing, Holmes and her crony are facing multiple charges of fraud and are battling to stave off jail time.

Most stories consist of all sorts of characters - good, bad, endearing and hateful. This story is not like that. Almost every person you will read about in this book is a fraud or behaves in an extremely stupid manner even when confronted with facts. There are no winners in this, just victims and the sociopathic perpetrators of fraud who still have the gall to go around behaving as if they are visionaries.

Read now on Amazon

DESCRIPTION 13

INTRODUCTION..17

CHAPTER 1: PRIPYAT AND CHERNOBYL BEFORE APRIL 1986 22

    THE CITY OF PRIPYAT..22

    THE CHERNOBYL NUCLEAR POWER PLANT 24

CHAPTER 2: DESIGN OF THE RBMK-1000 REACTOR 26

    HISTORY OF THE RBMK-1000 REACTOR 26

    DESIGN OF THE RBMK-1000 REACTOR 28

        The Pit, Vessel, Shielding, and Moderator *28*

        The Fuel Channels *32*

        The Fuel..*33*

        The Control Rods...*35*

        The Gas Circuit..*36*

        The Steam and Cooling Systems *37*

        The Emergency Core Cooling System (ECCS) *39*

        Confinement Of the Reactor...*40*

        Void Coefficient..*42*

    FUNCTIONING OF THE RBMK-1000 REACTOR....................44

        Summary of how the RBMK-1000 Works..............................*45*

        RBMK-1000 Reactors at the Chernobyl Nuclear Power Plant...*48*

CHAPTER 3: THE TEST OF REACTOR NUMBER 4 .50

The Intended Procedure 52
The Failed Afternoon Test 55
The Midnight Test 56

CHAPTER 4: THE EXPLOSIONS 60

The AZ-5 Button 61
The First Explosion 63
The Second Explosion 64

CHAPTER 5: EVENTS FOLLOWING THE ACCIDENT 68

The Graphite Fire 68
The Corium Lava 72
The China Syndrome 74
Immediate Deaths 76
Radioactive Leakage and Deposition 78
The Liquidators 82
Removal of Contaminated Debris 90

CHAPTER 6: CONSEQUENCES OF THE ACCIDENT 93

Effects on Human Health 93
Premature Aging *93*

- *Cancer* *94*
- *Nervous System Damage* *96*
- *Psychological Complications* *97*
- *Genetic Mutation* *98*
- *Infant Mortality Rates* *99*
- *Thyroid Diseases* *101*
- *Death Toll* *102*

EFFECTS ON THE ENVIRONMENT 107
- *Deposition of Radioactive Material* *107*
- *Urban Life* *108*
- *Forests* *109*
- *Herbs* *110*
- *Aquatic Life* *112*
- *Insects* *113*
- *Birds* *114*
- *Mammals* *115*

EVACUATION 116

ECONOMIC EFFECTS 119
- *Agriculture* *120*
- *Resettlement* *122*
- *Social Welfare* *122*
- *Direct Costs* *123*

PSYCHOSOCIAL IMPACTS 125

Trauma *125*

Distorted Demographics *125*

Psychological Distress *126*

Fear *127*

CHAPTER 6: INVESTIGATIONS AND TRIAL 129

THE INVESTIGATION 129

LEGASOV'S REPORT 131

THE TRIAL 133

Accusations *135*

Verdict *136*

LEGASOV'S DEATH 137

CHAPTER 7: CAUSES OF THE ACCIDENT AND RECTIFICATIONS 139

DESIGN FLAWS 140

Positive Void Coefficient *140*

Control Rod Design *141*

Containment *142*

HUMAN ERROR 144

Semi-Skilled and Uninformed Operators *144*

Low Power Operation *144*

Blocking the Automatic Control System *145*

Turning the Warning System Off *147*

 Turning the ECCS Off **147**
 Connecting All the Coolant Pumps to the Reactor **148**
 Removing All the Control Rods **149**
 Initiating the Test at Low Power **149**
 Pressing the AZ-5 Button **150**
RECTIFICATIONS OF THE RBMK-1000 REACTOR **152**
 The Control Rods **152**
 The Positive Void Coefficient **153**
 Reactor Shutdown Time **154**

CHAPTER 8: MODERN-DAY CHERNOBYL **156**

CLOSURE OF THE PLANT **156**
THE CHERNOBYL NEW SAFE CONTAINMENT **158**
TOURISM **161**

CONCLUSION **164**

Please observe a moment of silence for all that has been lost in the disaster.

Description

Have you read a book about a true event that led to the evacuation of whole towns because of something they could not forsee, never to return? Well, in 1986, it happened!

If you love reading books about stories and events that make your jaw drop and your eyes well up, this is exactly where you should be. The book whose summary you are reading takes you to the 1986 Chernobyl Nuclear Power Plant in the Soviet Union. One fateful night, what the people of the region refer to as the "peaceful atom" turned against them and wreaked havoc on their lives forever. In a few days' time, they would find themselves boarding hundreds of buses to escape from an invisible enemy that made their skin blister and their insides rot. This enemy would pursue them years after chasing them from their homes, only to cause serious illnesses to their bodies. The enemy's name is Radiation.

You need to download this book to read the story of Chernobyl and how just a routine safety check would result in the biggest meltdown that mankind has ever encountered. Everything that there is to know about the disaster is inside this book. You will not need to purchase a series of "Chernobyl" books to understand what happened.

Grab a copy to access hair-raising information such as:

- The history of the town of Chernobyl before the accident
- The design of the powerful reactor that caused the meltdown
- The nature of the simple test which turned fatal
- Step-by-step details on the events that caused an explosion
- The events that followed after the accident
- The consequences of the meltdown to humans, animals, and the environment

- Present life in the ghostly exclusion zone

If you love reading historical events that sound too scary to be true, then this is the perfect book for you. After reading it, anything else that you come across about the Chernobyl disaster will only be repetition.

See you on the inside!

Introduction

Congratulations on purchasing *The History Of Mankind's Greatest Disaster: A Walk Through The Chernobyl Nuclear Catastrophe*, and thank you for doing so. For anyone that knows anything about the name "Chernobyl", they probably understand the fear, sadness, uncertainty, and controversy that is associated with it. This name has inspired tons of books, movies, and even songs owing to the disaster that occurred in 1986, almost bringing the town of Chernobyl and the entire country of Ukraine to a standstill. Prior to that day, Chernobyl was just another power-generating plant in the town of Pripyat. Today, it is the largest reminder to humankind of the might of nuclear power, and the potential risks that it carries, if not granted the respect that it demands. A lot has been written about the disaster, but no one book has put everything that happened before and after in a single place. This book will do exactly that. We shall discuss everything that existed in Pripyat before April 26, 1986—everything that led to the meltdown on the fateful day, and life after the disaster. By the end of the book, the reader will have a clear and accurate flow of the events, as well as an in-depth understanding of Chernobyl, in terms of what led to man's greatest disaster, plus the effects that it brought.

Chapter 1 will provide a brief history of the city of Pripyat and Chernobyl before the disaster. It will show how life was normal and peaceful for the occupants who had been in the area since it was commissioned. The history of the town and city is important in the understanding of the role of the plant, as well as the events that led to the catastrophe in April 1986.

Chapter 2 will go deep into the design and functioning of the RMBK nuclear reactor used at the Chernobyl Nuclear Power Plant. Understanding how the reactor was made and how it was supposed to work is critical in understanding the flaws and errors that led to the failure, which led to one of mankind's worst disasters.

In chapter 3, we shall look at the nature of the test, which later developed into a disaster. The Chernobyl plant had no malfunctions up to the day of the explosions. The cause of the meltdown was a test gone wrong. In this part of the book, we are going to unearth the exact test that was going to be done, and the procedures that were supposed to be followed. It is only then that we can understand how a deliberate test could become a catastrophe.

Chapter 4 will describe the peak of the Chernobyl nuclear disaster: The accident. The accident was a result of a combination of operations by the personnel who were running the test on the fateful night. In this chapter, the reader will get insight into the step-by-step actions which put the reactor at strain, and led to two explosions that completely destroyed the Reactor Number Four.

Chapter 5 is dedicated to the events that occurred after the accident happened. These include the death of plant personnel due to the explosion, the emergence of a stubborn graphite fire, the risk of a third powerful explosion, efforts to clean the towns and so on. The events paint a scene of devastation and a lot of sacrifice, as the people united in trying to overcome one of mankind's greatest disasters.

Chapter 6 is about the consequences of the catastrophe. In this chapter, we shall look at all the ways in which normal life was interrupted. Apart from humans, the lives of plants and animals were also impacted. The environment, too, underwent significant changes. All these will be expounded upon inside this chapter.

The investigations into the disaster and the trials that followed the findings will be highlighted in chapter 6. This section will focus mostly on the chief investigator, as well as his findings, and the convictions of the defendants.

Chapter 7 will talk about the verified causes which led to the meltdown. Immediately after the disaster, different events were blamed, but the truth would come out about five years later. This chapter will also reveal the improvements that were made to the active reactors that remained at Chernobyl, and across the USSR.

In the final chapter, we are going to look at the status of Chernobyl, 30 years later. The old shelter for Reactor Number Four has been replaced. Interestingly, life seems appears to be coming back, only that, this time, the dominant creatures are not humans.

There are plenty of books on this subject in the market, so thanks again for choosing this one. Every effort was made to ensure it is full of as much useful information as possible. Please, enjoy!

☐

Chapter 1: Pripyat And Chernobyl Before April 1986

The City of Pripyat

Today, if you walk into the city of Pripyat, one of the outstanding landmarks that will greet you is a stagnant, old, rusty Ferris wheel that has not run for decades. Nature has taken over it, with birds and bats being the dwellers, and overgrowth covering its base. The saddest bit about this wheel is that it had never been used because its commissioning was planned for 5 days after April 26, 1986. Any surviving wall clocks in the abandoned houses are all frozen at 23:55, the approximate time when the explosion occurred and cut off all the power. This is the same case for everything in and around the now-abandoned town, and will remain like this for about 300 years.

Pripyat city was created at the beginning of 1970 in Northern Ukraine, which was then part of the Soviet Union. It was set near the Belarus-Ukraine border.

The name was borrowed from the Pripyat River, which flows along the city's border. Pripyat was founded as one of about a dozen nuclear cities or "atomic towns" in the Soviet Union, to serve as a dwelling for workers of the Chernobyl Nuclear Power Plant, which was just 3 kilometers away.

The town had been designed for a population of about 75,000 people. According to a 1985 census, the population then was 49,000. The inhabitants were mostly young people, who had come from all over the Soviet Union during the construction of the plant. Children comprised about one-third of the population. There were about 15 primary schools, with 5,000 students, 10 gyms, 25 shopping stores, a large hospital complex, an amusement park, small factories, cinemas, parks, and all other facilities that decorated a luxurious city in the 1980s.

The Chernobyl Nuclear Power Plant

Before it became the home of a nuclear power plant, Chernobyl, whose name translates to "wormwood", still had a rich history. It had strong Jewish influences from the beginning of the 16th century. When the Second World War arrived, Chernobyl was one of the most affected towns, and the Soviet Union made it their base for repairing their ships, owing to the vast River Pripyat, which flowed along the town's border. After the war, the USSR Ministry of Energetics chose the town as the site for a nuclear power plant, owing to its ready access to water and a low population.

The Chernobyl Nuclear Power Plant was constructed in the City of Chernobyl, about 100 kilometers from Kiev, the capital city of Ukraine, in honor of Vladimir Ilyich Ulyanov Lenin. Lenin was a theorist, politician, and revolutionary, who was the head of government in Soviet Russia between 1917 and 1922. He died in 1924. The plant's construction begun in 1970, at the same time as the town of Pripyat, and the first reactor was completed and commissioned in 1977. It became the first nuclear power plant in Ukraine. It would comprise of 4 nuclear reactors, which would collectively provide 10% of Ukraine's electricity. The initial plan was to make the Chernobyl Power plant the largest nuclear power plant in the world. It would have 12 reactors, and produce 12,000 megawatts of power. However, by the time of the disaster, only 4 reactors were operational, and 2 more were under construction.

Life was smooth since the town and plant were commissioned. There had been no major incidences in the plant prior to 1986. Normally, there were two shifts: One for the day, and another for the night. As the shifts changed, the outgoing head engineer would brief their incoming colleague to ensure that operations were smooth and safe until the next shift. This was the norm until the early morning of April 16, 1986.

Chapter 2: Design of The RBMK-1000 Reactor

History of the RBMK-1000 Reactor

The famous RBMK-1000 reactor derives its name from the Russian phrase, *Reaktor Bolshoy Moshchnosti Kanalnyy*, which translates to "a reactor that uses a high-power channel." The name is derived from its design, which uses individual fuel channels enclosed in pipes, instead of huge steel pressure vessels used in other reactors. This was a unique type of nuclear reactor which used graphite as a moderator, light water as a coolant, and enriched uranium as fuel, and which was designed by the Soviet Union. Light water is normal water with minimal amounts of deuterium, and is highly preferred in electricity generation, as it can act as both a coolant and moderator in carrying away the energy released during a nuclear fission. Initially, the RBMK was made for the production of plutonium, but was later widely adopted in the generation of electricity. The combination of light water as a coolant, and graphite as the moderator made it a reactor of its own, as no other reactor in the world combines both. A single reactor of this type had the capacity to produce 1,000 megawatts of power for approximately 45 years.

Design of the RBMK-1000 Reactor

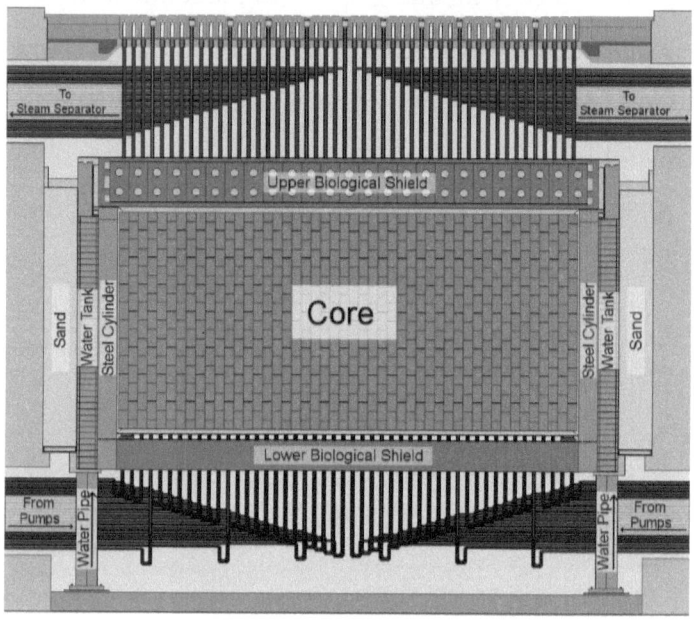

Fig. 1: A cross-section of the RBMK reactor's structure. Photo courtesy of *Wikimedia*

The Pit, Vessel, Shielding, and Moderator

The pit is the outermost part of the reactor, which holds the vessel in position. It is made using reinforced concrete, and measures about 21.6 meters by 25.5 meters by 21.6 meters. The pit holds the vessel, which is the second outermost layer of the reactor. It has a steel cylindrical wall and metal plates on the bottom and upper sides. It measures 14.5 meters in height, 9.7 meters in width, and its walls are 1.6 centimeters thick. The vessel holds the stack of graphite, and is full of a mixture of helium and nitrogen. This mixture creates an inactive environment to allow for the transfer of heat from the graphite stack to the coolant channels.

Inside the vessel are moderator blocks, which are made out of nuclear graphite. They measure 25 centimeters by 25 centimeters. Inside the longitudinal axis of the blocks, there are 11.4-centimeter holes that hold the fuel and control the rod channels.

Around the moderator is a cylindrical water tank with an inner diameter of 16 meters and an outer diameter of 19 meters. The walls are 3 centimeters in thickness. This tank is divided into 16 vertical compartments, which are fed with water from the bottom, which exits from the top. This water is meant for emergency cooling of the reactor. Inside the tank are thermocouples for sensing the temperature of the water, and some ion chambers, which monitor the power of the reactor. This tank, plus a layer of sand and the reinforced concrete, act as biological shields to keep radiation from getting outside of the reactor.

On the reactor's top, there is a cylindrical disc called the "Upper Biological Shield", or UBS. The disc measures 3 meters in height and 17 meters in diameter. The disk is penetrated by the fuel and control channel standpipes. The top and bottom ends of these channels are sealed using thick plates made of 4-centimeter-thick steel. The gap between the pipes and the plates is full of serpentinite. This is a rock that contains bound water—that is, water that is physically or chemically attached to the rock such that it cannot be removed without changing its natural structure and composition.

The UBS is covered with an upper shield cover, which contains individual steel-graphite plugs that can be removed. These are placed in the center area of the cover, which is directly above the reactor channel. This cover acts as thermal insulation, an additional biological shield, and as the floor of the central hall which holds the reactor pit.

On the lower side of the reactor core is another biological shield known as the "Lower Biological Shield", or LBS. It is identical to the UPS, only differing in that it is only 2 meters in height. It is penetrated by the lower ends of the pressure channels, and also carries the weight of the coolant inlet piping and the graphite stack. There are two heavy pipes that intersect below the LBS, and are welded to it, and whose task is to support it and transfer the load of the entire reactor into the building.

The Fuel Channels

The network of the fuel channels is made up of 8-centimeter-wide zircaloy pressure tubes, which are 4 millimeters thick. These pass through the channels in the center of the moderator (graphite) blocks. The top and bottom sections of the tubes are made of stainless steel, and the zircaloy central segment with couplings is made of zirconium-steel alloy. The pressure tubes are held in the channels within the graphite stack, using two high-split graphite rings. One of the rings directly touches the tube, while the other touches the graphite stack. This assembly allows for heat transfer from the graphite blocks.

Most of the heat from the nuclear fission is generated in the fuel rods. However, about 5.5% is transferred to the graphite blocks during the moderation of the fast neutrons, which are created by the fission. This heat should be removed to prevent an overheating of the graphite stack. Most of the heat energy transferred to the graphite is discarded by the fuel rod coolant channels through the graphite rings. The remaining heat is discarded from the rod channels using forced gas circulation.

First-generation RBMK-1000 reactor cores have 170 control rod channels and 1693 fuel channels. The second-generation reactors like Chernobyl's reactors three and four have 211 control rod channels and 1661 fuel channels.

The Fuel

The reactor uses fuel pellets of uranium dioxide powder which has been compacted using a binder to form a solid mass inside barrels that are 11.5 millimeters wide in diameters and 15 millimeters in length. Europium oxide might be added to the uranium dioxide to form a burnable nuclear poison, which helps in lowering the reactivity differences, which may occur between a partially spent and a brand new fuel assembly. The pellets have some hemispherical indentations, which help to reduce issues with thermal expansion. There is also a 2-millimeter bore in the middle of the pellets, which reduces the temperature at the center, and allows for the expulsion of gaseous fission by-products.

The fuel rods are made of zircaloy. They are 0.8-millimeters thick and 13.6 millimeters in outer diameter. Helium is put inside the rods before they are hermetically shut. There are retaining rings that keep the pellets in the intended position, and also enable the transfer of heat from the pellets to the tubes. A single rod has approximately 3.5 kilograms of pellets. The rods measure 3.6 meters in length, although the indentations consume about 0.2 meters of the active zone.

At any single time, the full capacity of a reactor in a stationary condition is about 192 tons of radioactive fuel.

The Control Rods

Boron is used to make the control rods for the RBMK-1000 reactor. These rods are used to control the reactivity of the uranium neutrons. When the rods are inserted, they reduce the fission of the uranium fuel, thus lowering the reactivity and power output of the reactor. When they are retracted, fission increases, thus there is higher reactivity and more power output. All but a few control rods are inserted from the top of the reactor. There are 24 shortened control rods, which are inserted from the bottom to augment the distribution of axial power in the reactor's core. The control rods have 4.5 meter-long graphite tips that are separated using a boron carbide neutron absorber, and a 1.2 meter-long telescope, which fills with water to create a space between the absorber and the graphite. The graphite, also known as the "displacer", is used to boost the difference between the attenuation levels of the neutron flux between the inserted and retracted rods. When a control rod is fully removed, the graphite is positioned in the middle of the core, cushioned by water on either side. As the water gets displaced when the rod is moving down, it increases reactivity at the bottom of the reactor core as the graphite tip passes through that area.

The cooling of the control rod channels is done by an independent water circuit system, which maintains the temperature between 40 and 70 degrees Celsius. The space between the channel and the rod slows the flow of water around the rods when they are moving, thereby acting as fluid dampers. This is why they are inserted slowly.

Normally, only 43 to 48 control rods are required to maintain the optimum reactivity margin.

The Gas Circuit

The RBMK-1000 reactor contains a gas circuit used to create the helium-nitrogen atmosphere that is necessary for its functioning. The mixture is made up of 10 to 30% nitrogen and 70-90% helium. The circuit is made up of an electric compressor, iodine and aerosol filters, an absorber for ammonia, carbon monoxide, and carbon dioxide, a tank for holding the gaseous byproducts of nuclear fission until they decay before being released, a ventilator stack, an aerosol filter to discharge solid decayed products, and a chimney to release safe gases into the atmosphere. The gas is driven into the stack from the lower end at a low rate, and comes out through the standpipes of every channel through individual pipes. The temperature and moisture levels of the gas being discharged are constantly being monitored, as an increase in these two could indicate leakage of the coolant.

The Steam and Cooling Systems

The RBMK-1000 reactor has two cooling circuits, which operate individually. Each circuit has four pumps, where three are constantly operating, as one is on standby for emergencies. The water for cooling is pumped into the reactor using lower water lines, where they meet common pressure headers, one for each circuit. The headers are split into about 22 distribution headers, where each of them feeds about 40 pressure chambers where the feedwater boils.

When the water boils, the mixture of water and steam is directed by the upper steam lines from the top of the reactor to the steam separators, which are thick drums positioned inside compartments above the top of the reactor. The water settles at the bottom of the tanks as the steam rises to the top, where it is collected and combined with steam from the other separator. From here, it is led to the turbine hall, where there are two turbogenerators. When it leaves the turbines, it is led to a condenser, gets reheated to about 165 degrees Celsius, and gets pumped to the deaerators. Here, the remains of corrosive and waste gases are removed. The feedwater that remains is led down to the main circulation pipes, where it once again finds its way to the reactor.

The main pumps are electrically powered by 6-kilovolt motors to provide a capacity flow of between

5,500 to 12,000 cubic meters of water per hour. This capacity is slowed down to about 6,000 to 7,000 cubic meters per hour by control valves whenever the power of the reactor goes below 500 megawatts thermal (MWt).

For safe operation of the reactor and plant at large, the amount of steam in the reactor tubes, the level of water in the separator tanks, the distribution of power in the reactor, the neutron flux, the level at which the water starts boiling in the reactor, and the rate of flow of the feedwater in the core must be carefully observed and maintained.

The Emergency Core Cooling System (ECCS)

There is an ECCS in the reactor which is made up of pumps, hydraulic accumulators, and dedicated water reserve tanks. This system flows along the dedicated reactor cooling system described above. It is used to keep the turbogenerators running in case of a total loss of power to the nuclear plant, until the backup diesel generators kick in and maintain the functioning of the reactor. The ECCS comprises three systems: The first one provides water to the affected half of the coolant system immediately if the power is cut off for at least 100 seconds. Then, the other two pick up from there, and provide long-term cooling support.

These pumps are powered by 6-kilovolt internal power lines, as well as diesel generators for backup. There are some valves that require a constant power supply, and these are backed up using batteries.

Confinement Of the Reactor

The RBMK-1000 reactor was preferred at Chernobyl because it was powerful, yet quick to build and easy to manage. However, in terms of containment, there was a worrying omission, in that the reactor did not have full physical containment structures on the top. One reason is that the reactor was too tall, and constructing a containment structure to cover it to the top would have cost more and taken more time. Additionally, the fact that the fuel rods could be changed when the reactor was at full operational power meant that cranes would be required for this task. As such, containing the cranes inside a structure would mean even more design constraints. Therefore, the upper containment structure of this reactor was below the recommended strength.

The bottom part of the RBMK reactor is positioned inside a watertight compartment and there is a gap between the floor and the reactor's bottom. There is a Steam Distribution Corridor under the reactor, whose task is to contain the steam in the event of an accident. If any of the pressure tubes ruptured, the steam would go down this corridor and bubble through the water in the compartment so it condenses and reduces the pressure in the tubes without letting the steam leak. This Steam Distribution Corridor is made up of a fire sprinkler system, jet coolers and surface condensers, which are used to remove radioactive aerosol molecules, cool the air in the reactor, and remove steam.

Void Coefficient

The void coefficient of the RBMK-1000 reactor is positive. This reactor is cooled by water. Therefore, when it gets too hot and the water boils, some steam bubbles are produced in the coolant or moderator. Water is a more effective coolant, as well as a neutron absorber, as compared to steam. Steam reduces the ability of the water to absorb the unwanted neutrons, as well as cools the reactor, allowing more neutrons to reach the graphite. When the graphite shied reflects them back, fission increases. As such, when more steam is produced, the void in the coolant increases reactivity in the reactor's core, leading to the production of more power, thus more heat, and even more steam. If the reactor had a negative void coefficient, the reactivity in the core would reduce as steam bubbles formed. The RBMK reactor has an extreme positive coefficient, meaning, it has the ability to rapidly become unsafe. □

Functioning of the RBMK-1000 Reactor

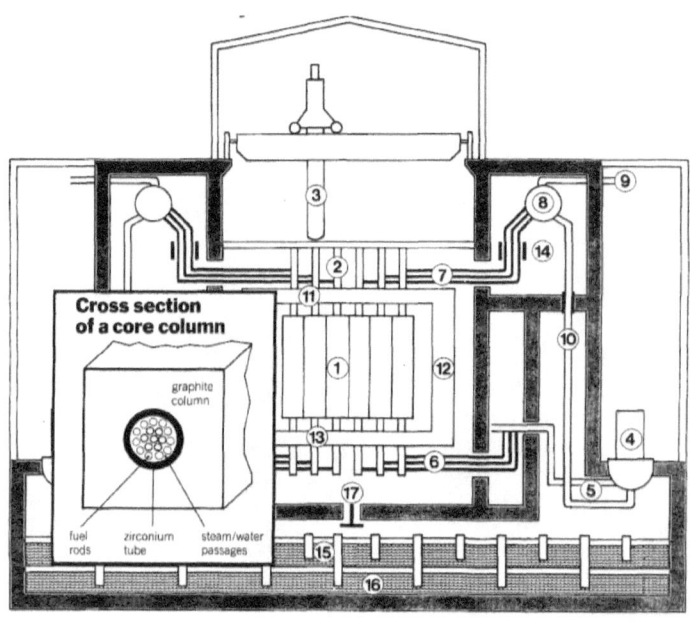

Fig. 2: A graphical representation of the RBMK reactor. Photo courtesy of *Bulletin of the Atomic Scientists, 1986*.

Summary of how the RBMK-1000 Works

1: In the center of the reactor, the core graphite blocks are used as the moderator of the nuclear fission reaction. The graphite blocks have channels passing through them to accommodate the fuel and control rods.

2: Boron control rods are lowered or retracted from within the graphite blocks to decrease, increase, or stop the nuclear reaction.

3: A crane placed above the reactor is used to insert rods of enriched uranium (fuel) inside fuel channels within the graphite blocks. These fuel rods are the source of the nuclear fission, which releases heat as the uranium atoms split.

4: Main circulation pumps send water throughout the reactor. These are fed with external power, and can be powered by diesel generators in case of power loss. In the core, the water draws heat from the reactor, and some of it boils to form steam.

5: Multiple distribution pipes direct the water from the pump to the core to control the temperature and prevent overheating of the reactor, or the formation of excess steam.

6: Narrow coolant pipes subdivide from the main water distribution pipes and attach to the fuel channels. They draw heat from the channels through conduction, and the water absorbs the heat.

7: At the top of the reactor, the hot water and steam, which has passed through the reactor's core, are collected by the coolant outlet pipes.

8: The coolant outlet pipes direct the water and steam mixture into huge tanks that separate the water and the steam.

9: The steam rises to the top of the tank, and is led out by pipes at high pressure, and it is used to turn two turbogenerators that produce electricity.

10: Once the steam turns the turbines, it is led out through pipes that direct it to the main circulation pumps, where it condensed back into water, and rejoins the main circulation systems once again. From here, the process begins again.

11, 12, and 13: Multiple biological shields are placed in strategic points within the RBMK-1000 reactor. The shields protect plant workers from the radiation present in the core of the reactor. The spaces within the shields are full of inert gases.

14: The cladding failure detection system acts as extra security to detect any radiation leaks, which might be present in the coolant or anywhere in the reactor.

15 and 16: At the bottom of the reactor, there are bubbler pools that would receive any leaked steam, and condense it to prevent it from escaping.

17: Finally, any steam that might be released within the reactor compartment is driven out through pressure relief valves, and condensed into liquid in the bubbler pools.

RBMK-1000 Reactors at the Chernobyl Nuclear Power Plant

The first RBMK reactor at the Chernobyl Nuclear Power Plant was completed in 1977. The second one went live in 1978, the third in 1981, and the last one in 1983. They were named "Reactor Number One", "Reactor Number Two", "Reactor Number Three", and "Reactor Number Four" respectively. There was a slight difference in the design of reactors numbered three and four. Unlike one and two, which were first-generation RBMK reactors, three and four were second-generation. This means they had a better containment design and structure.

The reactors numbered five and six, following the same design, were planned on a site that was about 1 kilometer from the building that housed the first four reactors. As at the time of the explosion, Reactor Number Five was over 70% done, and would have been ready in about 6 months. The construction of the two reactors was suspended following the explosion, and was later canceled in April 1989. This was a few days to the third anniversary of the 1986 disaster. The other six reactors, which would have been constructed on the opposite side of the Pripyat River, never saw the light of day.

Chapter 3: The Test Of Reactor Number 4

The Chernobyl Nuclear Power Plant had been running for many years since its commissioning, and there had been no incidents affecting any of the reactors before 1986. The plants had proven efficient and safe when handled, as recommended by the chief engineers who had designed them. However, there was one test that had been giving the engineers at the plant sleepless nights: The test to see whether the spinning turbochargers would keep the cooling system in working condition in the event that the entire plant lost power.

When a nuclear reactor is in full operation, there is a significant portion of power, about 6%, which comes from byproducts of the nuclear fission. As the accumulated products decay, they give off significant heat, which is also used in heating the water and giving off steam. As such, even after the chain reaction has been stopped by the insertion of the boron control rods, some heat is still generated by these products before they decay fully. Therefore, some cooling is still required to keep the reactor safe until the products stop producing heat and keep the reactor core from melting down.

There were three backup diesel generators that would keep the water pumps working when the power went out, but they required between 60 to 75 seconds before they attained full power. This short period of time is very critical, because the lack of a cooling mechanism in the reactor for that long would make it unstable and highly risky. It had been theorized that the system had the capacity to bridge this critical 1-minute delay using the stored rotational inertia, which remained in the turbines when the power went off. Concisely, they thought that the time which a spinning turbine took, from the moment the power was cut until it came to a halt, would provide enough electricity to the pumps before the generators had enough power to take over. It was estimated that the turbines could produce power for about 45 seconds before stopping.

This theory needed to be put to a physical test. The first test was carried out on Reactor Number Four in 1982, but it failed. A few modifications were made, and the test was repeated again in 1984, but it failed yet again. They tried again in 1985, and, to their disappointment, the experiment was still unsuccessful. After a year of modifications, the engineers felt they were ready to conduct a successful test on April 25 1986, during the day shift.

The Intended Procedure

The idea of the safety test was to imitate an emergency electricity shutdown, which would require the switching of the sequences of the electrical supply to the reactor. Rather than use power from the grid to run the coolant pumps, the engineers hoped to test whether the remnant power in the spinning turbines could maintain the pumps long enough until the diesel backup generators came online. Since it was going to be a deliberate emergency situation, the scientific manager of the plant and the chief designer of the RBMK-1000 reactor were not required to be present. The test would be under the supervision of the plant director.

On the day of the test, the director responsible for Reactor Number Four, Victor Bryukhanov, happened to be out of town. The responsibility for the test was therefore handed to the deputy chief engineer of the Chernobyl Power Plant, Anatoly Dyatlov. Dyatlov had never supervised any experiment of that nature before.

The test conditions required that the total thermal output of the reactor be a minimum of 700 megawatts—about half the normal total output—during the start of the test. On the 25th of April 1985, the power reduction began at 1am. Dyatlov was overseeing the process, and a foreman, Aleksandr

Akimov, plus a few other men, were reporting to him. The power reduction had to be gradual, due to the production of xenon, a by-product of the nuclear fission process, which absorbs neutrons without splitting. If the reduction was performed too fast, a lot of xenon would be produced, and it would absorb a lot of neutrons, leading to the stalling of the fission process. In this state, the core could explode. On the other hand, if the power was gradually reduced, it would allow the xenon gas to decay without affecting the nuclear fission process.

When the power had reduced to half, after about 12 hours, one of the turbines driven by the reactor under test was shut down because the available power could only sustain one turbine. After it was shut down, power would further be reduced to 30%, and the remaining turbine would be shut off. At this point, the engineers would observe how long it took the blades of the turbine to come to a complete stop. If the time was enough to keep the coolant pumps running until the generators were ready, the test would be considered a success, and everything would be restored to normal.

The Failed Afternoon Test

The test was ready to begin on the afternoon of April 25 1986. However, just as Dyatlov was about to begin the test, he was called by a Ministry of Energy official from Kiev, and was told that a grid had lost power in the area, so they were not to lower the power further. As such, they had to postpone the experiment until the next shift. The emergency alert system and the Emergency Core Cooling System, ECCS, had both been shut down, yet, after the test was postponed, they were not restored. Similarly, the reactor was left to run at the low power state—a situation that was known to make the RBMK-1000 reactor unstable.

The Midnight Test

At 11.04 pm, the Kiev official informed Dyatlov that the shutdown could resume. This was about an hour before the day shift ended, and the night shift workers took over. The day shift workers had been prepared for the main test, while the night shift personnel had only been prepared to maintain the cooling of the reactor, which would have been shut down before their shift began. A rapid power reduction was initiated as Dyatlov prepared to start the test when the night shift began. Shift supervisor Aleksandr Akimov was present, and the operator at the controls was young engineer Leonid Toptunov. At five minutes past midnight, the power got to the level required by the deputy chief engineer. The test was about to begin.

The engineers in the control room realized that, despite stopping further their lowering of power, the level was still decreasing. The rapid reduction of power had allowed xenon to build up and absorb more neutrons than necessary. This had caused the nuclear fission process to die down in a process called "reactor poisoning". Further absorption of the neutrons took the power lower to 500 megawatts, and then it suddenly slumped down to about 30 megawatts.

30 megawatts represented about 5% of the total power output of the reactor, yet a minimum of 700 was required for the test. It would have been mandatory to shut down the system and abandon the test, but Dyatlov was unwilling to fail, and too impatient to allow for better preparation. He ordered the young engineer to remove seven out of about 18 boron control rods in the reactor core to allow for fission to resume and power to rise, so that they could resume the experiment.

Since the reactor had been operating at reduced power since the afternoon, the water pumps were also slow. Between 12.35 am to 12.45 am, the high poisoning levels, low steam levels in the separating drums, irregular coolant flow, the opening of relief valves to allow excess steam into the bubbling tanks, and unstable temperature in the reactor core triggered warning alarms, but the duty personnel ignored them.

The removal of the control rods seemed to work a little, because power rose to about 200 megawatts. Convinced that the reactor was back to normal, Dyatlov instructed his subordinates to resume the test. More water was pumped into the core, leading to an increase in the temperature of the inlet coolant, since the increased speed meant the coolant did not have enough time to cool down. The water gradually approached its boiling point, causing a reduction in the safety margin of the reactor.

At about 1.19 am, an alarm for low steam pressure in separator drums went off. Simultaneously, the increased water flow lowered the temperature and reactivity of the core, thus reduced the power output. The engineers tried to resolve the stagnation in the low power output by disabling two circulation pumps to boost the steam pressure. They also removed more manual boron control rods to increase reactivity. This combination of operations made the reactor unstable. Almost all the 211 control rods had been removed, leaving only 18. The minimum recommended number of inserted rods was 28. An emergency scram system that was meant to take over when the reactor was in this state had been disabled, leaving the entire control to manual operation. The system could still be activated manually by engaging the AZ-5 button, but this did not happen. When the coolant flow was reduced, the safety margin fell further, and any further power increase the risk of boiling all the water in the cooling channels. The reactor was on the brink of disaster, but no one in the room seemed to be aware of it.

Chapter 4: The Explosions

The test seemed to be going according to plan as at 1.23 am. The steam feed to the turbogenerators was switched off, and the turbines began the gradual deceleration. The standby generators turned on as expected, and were expected to have been at full power by 1.24 am. During the exchange gap that occurred, the slowing turbines reduced the power that the main circulating pumps were receiving. This, in turn, decreased the amount of water flowing in the cooling system, and subsequently, caused the formation of steam voids in the fuel pressure tubes, which flowed towards the top of the reactor.

As mentioned earlier, the void coefficient of the RBMK-1000 reactor is positive during a low-power operation, unlike most other reactors. As such, when the coolant boiled excessively inside the fuel pressure tubes, it produced large steam voids in it, rather than bubbles. The formation of the voids, in turn, increased the chain reaction, because it reduced the relative amount of water required to absorb the neutrons. When the power increased, more voids were formed, and this further catalyzed the chain reaction. At this point, Chernobyl's Reactor Number 4 was out of control, and there was nothing to slow it down.

The AZ-5 Button

Toptunov realized that the power was doubling every second, and that the reactor was heating up rapidly. He informed Dyatlov, who, upon realizing the danger that was at hand, instructed Toptunov to initiate the emergency shutdown, popularly known as the "SCRAM". The time now was at 1.23.40 am. To initiate the SCRAM, a button known as the "AZ-5" or "EPS-5" on the control panel had to be pressed. When this button is pressed, the first action it takes is to stop reactivity in the core, by inserting all the boron control rods. A recording provided by the computer system showed that the AZ-5 was manually pressed, although there is no evidence as to why it was engaged.

The boron rods, 7 meters long, dropped into the reactor core at a speed of 40 centimeters per second. Therefore, it took approximately 18 seconds for them to go all the way in. This speed was too slow, and it did not help to cut down the reactivity. Another problem presented itself at this critical moment: The tips of the control rods were made of graphite, which was meant to enhance the output of the reactor by displacing the coolant (water), as the active section of the rods was fully withdrawn from the reactor. Now, as the rods went down into the reactor, they displaced the water used in absorbing the neutrons, and the graphite, which moderated the neutrons, replaced it. The immediate response of this action was increasing the reaction rate in the lower end of the RBMK-1000 reactor.

The First Explosion

Power in the reactor doubled with every millisecond, and in just four seconds, the output peaked to 100 times the normal capacity of the reactor. This is approximately 30,000 megawatts. A nuclear burst caused by the rapid rise in temperature to about 3,000 degrees Celsius caused the coolant to turn into steam, and the resulting sudden pressure ruptured the pressure tubes. When the water gushed out and came into contact with the graphite moderators, there was an explosion. This explosion shattered the upper biological shield and the reactor casing, which weighed about 907 tons. More coolant lines and fuel channels were ruptured, causing the remaining coolant to come into contact with the graphite and escape.

The Second Explosion

The core of the reactor, now left without any coolant, produced more thermal power. This caused a second explosion, three seconds after the first, which was louder and more powerful. The second explosion scattered the damaged core and terminated the nuclear fission. It caused further shattering of the reactor, and threw hot graphite outwards to the damaged roof of the plant. As the graphite came into contact with the air, it burst into flames, and started huge fires on the roof and inside the remains of the reactor. It is estimated that over 25% of the graphite in the reactor was ejected.

There are several explanations offered as to what caused the second explosion. It is highly likely that the first explosion happened when rapid boiling caused water to turn into steam. Due to the high pressure that followed, the pressure tubes were severed, and the explosion was as a result of the steam spreading out to other parts of the reactor. One theory offered for the second explosion is that the uncontrolled escape of the neutrons caused the explosion, as water suddenly disappeared from the reaction chain as thermal energy peaked. A second theory is that, when the steam reacted with the hot graphite, hydrogen was produced, and it exploded. There are many theories in place which try to explain the occurrence, but none has ever been quantified as the exact cause.

The two powerful explosions sent the structural remains of the reactor and hot radioactive debris such as the remaining fuel and graphite into the night air, and exposed the shattered core to the atmosphere. The thick plume of radioactive smoke and the smoke from the burning debris rose well over 1 kilometers into the sky. Some witnesses reported seeing a glowing streak of white light emanating from the burning plant, illuminating the clouds above. The heavy debris was cast around the plant's compound and the roof of the remaining reactors. The lighter components, such as the noble gases and fission products, rose to the atmosphere, and were carried by the wind towards the North West of the Chernobyl Nuclear Power Plant.

Fire started on the roof of the remaining plant because it had been made of bitumen, which is combustible. The men in the control room were cast into pitch darkness, as the plant lost power. The walls of the hall that held the reactor were destroyed, and the pit that held the reactor was nothing but a hole full of burning debris. There was nothing left of the roof, and the core was now in the open air. The part of the building that remained had no glass remaining on the windows, and most of the walls and ceilings were cracked. The steel doors were twisted, and the survivors had to weave their way around the dark building, which was now full of smoke and steam.

This amount of destruction had taken place in less than a minute after the safety test had begun. Chernobyl's Reactor Number Four was no more.

Chapter 5: Events Following The Accident

The Graphite Fire

As the fire broke out in the damaged reactor and the roof of adjacent turbines and stores, the firefighters from Pripyat were called in, and the first group of 14 firemen arrived at 1.28 am. Additional reinforcements were called in from other areas, including the capital city, Kiev. In total, about 250 firemen were at the scene battling the fire. By 2.10 am, the emergency responders had managed to quell the flames on the roof of the machine hall. They intensified their efforts, and had put out the large flames on the roof of the reactor hall by 5 am.

At this time, the graphite fire was picking momentum, and it did not respond to the water used by the firemen. Unbeknownst to them, the new fire was the most dangerous of all, because it dispersed fission fragments and radionuclides, such as strontium-90, iodine-131, and caesium-137, around the area and into the atmosphere. The Soviet authorities had not disseminated any knowledge on fighting graphite fires, nor responding to nuclear disasters, since they deemed the reactors as "safe" and almost impossible to cause disaster. This turn of events exposed the men to the radioactive material and they took in huge doses. Ten of them would die of the poisoning.

By morning, word had spread out about the Chernobyl disaster, and nuclear experts ordered the evacuation of the firefighters, as they were endangering their lives. Unfortunately, this was too late, as a number of them had already received fatal doses of poison. The new approach taken to put out the graphite fire was to use huge amounts of specific materials, each to target different aspects of the fire. First, it was decided that the release of the harmful radionuclides should be the priority. Tons of neutron-absorbing materials were to be delivered to the reactor: 2,400 tons of lead, 40 tons of boron compounds, 600 tons of dolomite, 1,800 tons of sand

and clay, as well as polymer liquids and sodium phosphates were dumped into the reactor by May 2, 1986.

Helicopters with special containers were used to fly the materials over the reactor and pour it from above. At first, they would hover directly above the burning core and pour the material, but it became unsafe, since the pilots were exposed to high doses of radiation. Thereafter, the helicopters had to maneuver and pour the materials while in motion. This caused further destruction of the structure, and more contamination around the area occurred. The lead acted as an absorber for the radiation, the dolomite acted as a heat sink and producer of carbon dioxide, which would choke the fire, the boron acted as the neutron absorber, and the clay-sand mixture suppressed the release of particulates from the fire. On May 9, 1986, the graphite fire was finally extinguished, putting an end to two weeks of radiation leakage into the atmosphere.

The Corium Lava

While the firefighters battled the fire, and the world watched in horror as more poison was released into the sky, the scientists at Chernobyl were having sleepless nights over a third possible explosion, which would be even more powerful. It would not only destroy some of the remaining reactors, but also release more radioactive poison into the surrounding areas and into the sky. The molten material from the reactor, plus the smoldering graphite, was burning at over 1,200 degrees Celsius. Due to the high temperatures, the mixture started melting the concrete floor, forming corium. This is a form of semi-liquid radioactive lava. The engineers feared that, if the corium lava burned its way down to the bubbler tanks, which had millions of cubic meters of coolant, the resulting explosion would leave the entire site in shambles and endanger the lives of the people living in Pripyat, Kiev, and the entire European continent. The tanks were flooded due to the water used by the firefighters and coolant from the ruptured cooling pipes.

It was decided that they would not wait for the lava to get to the water. The scientists revealed that, if they could get to the bubbler pools underneath the plant, the sluice gates could be opened, and this would drain the tanks. Three volunteers decided to execute the suicidal mission. They were Valeri Bezpalov and Alexei Ananeko, both engineers at the plant, and who knew the location of the valves, accompanied by Boris Baranov, their shift engineer. They wore respirators and wetsuits to protect themselves from radioactive aerosols, and went into the flooded building with contaminated water up to their knees. Their mission was successful, as they found the valves and let out the water in the bubbler tanks. At this point, the risk of a third explosion was thought to have been eliminated—but they were wrong!

The China Syndrome

More information surfaced that, since the meltdown was still active, there was another risk for a third, more powerful explosion, known as the "China Syndrome". This is when molten material from a damaged nuclear reactor burns its way deep into the Earth. If the syndrome happened, and the corium lava reached the water table below the plant, there was a possibility of another steam explosion. In addition to the explosion, the corium could contaminate the Black Sea, which was a source of water for millions of people. Several suggestions were offered on how to freeze the ground below the plant so the corium lava would not bypass it. One of the feasible actions was to drill under the reactor, and inject tons of liquid nitrogen to freeze the soil, as well as stabilize the foundations of the plant. However, this plan was not put into action.

The alternative plan, and which was quickly adopted, was to deploy coal miners to dig a tunnel under the reactor for the installation of a cooling system. The system would comprise a coiled network of pipes that would be fed with cold water. They would be coated with a thin layer of conductive graphite, which would cool the corium lava when it came into contact with the pipes. The graphite would protect the pipes from being burnt. The whole network would be cushioned in two layers of 1-meter-thick concrete.

Just like the nitrogen plan, this one was not fully implemented. The miners dug the tunnel, and just before the cooling system was installed, the temperature of the corium lava dropped significantly. Investigations showed that, after burning through three floors, the lava had condensed in one of the basement rooms. Unfortunately, for the miners, about one out of every four had been exposed to fatal amounts of radiation in their generous efforts to save the world.

Immediate Deaths

Despite the immense scale of the two explosions, only two people died at the scene. When the reactor blew, 35-year-old Valeriy Khodemchuk, a pump operator, had been standing a few feet away from the reactor in the main circulating pump room. It is suspected that he was vaporized by the heat, as his remains have never been recovered. Normally, if a person was exposed to 100 roentgen (the units used to measure radiation) per hour, and for 5 hours, the dose was enough to kill them. Khodemchuk was in the reactor hall when the radiation probably spiked up to 30,000 roentgen per hour. The intensity of this radiation can be equated to 10 nuclear bombs, of the type that was dropped in Hiroshima.

The second fatality at the scene was Vladimir Shashenok, a 35-year-old plant worker, who happened to be in the instrument room adjacent to the reactor. Steam was suspected to have demolished the chamber, and he received serious scalding and radiation burns all over his body. His exposure was so high that the two colleagues who dragged him out of the chamber got burned from radiation in the areas where his skin came into contact with theirs. At the hospital, his whole body turned into blisters, and he died approximately 5 hours after arriving in the hospital.

Khodemchuk and Shashenok were the only people who died as a result of the two explosions. More deaths started being reported two weeks after the accident. One of the victims was Aleksandr Lelechenko, a 47-year-old plant electrician who went into the plant repeatedly to implement emergency repairs. 28 other deaths followed. Most of them were the first people who responded to the scene, such as 6 firefighters, 2 local police—one of whom was female—plant security guards, medical personnel, members of the Soviet Army, and plant workers who were exposed to excess radiation as they saved others and kept the remaining reactors safe.

Due to the high contamination in their bodies, the 31 bodies were enclosed in zinc coffins that were welded shut, and later covered with liquid concrete to prevent contamination of the environment.

Radioactive Leakage and Deposition

The leakage of radioactive material was the biggest threat of the Chernobyl Nuclear Power Plant's accident. Highly dangerous radioactive materials, consisting of finely fragment fuel, aerosols, and gases were released when the reactor's core exploded. All the gaseous elements, such as xenon and krypton in the fuel material, escaped into the atmosphere. Highly volatile materials, such as tellurium and caesium, attached themselves to the present aerosols, and were carried by the air. The extensive burning period of the graphite and remaining fuel enabled even the less volatile elements to be carried away. This was because, as the fuel burned, these materials bound to the hot fuel particles, which were lighter, and they were carried together by the wind. Some of these elements included strontium, zirconium, cerium, lanthanum, barium, and actinides.

Statistics showed that almost every country in the Northern Hemisphere was contaminated in a way—an occurrence that is attributed to the meteorological conditions at the time of the disaster. On the first ten days after the explosion, the weather changed frequently. This led to a wide dispersion radius of the radioactive material. The biggest particles were deposited within a 100-kilometer radius around the plant. The lighter particles traveled further North West, being carried in a cloud by the wind to other parts of Europe. Of the total radiation emitted by the cloud, 2% was caesium-134, 4% was caesium-137, 7% was barium-140, 36% tellurium-132, and 46% was iodine-131. Most of these substances were deposited with rainfall. However, the highest concentrations remained in South Belarus and Northern Ukraine.

There were three major contamination hotspots identified after the accident. They were:

1. The *Central Spot*, which was the most contaminated zone. It was the region within a 30-kilometer radius around the Chernobyl Power Plant. Here, the heaviest particles were deposited immediately after the blast, and more continued to fall until the graphite fire was extinguished.
2. The *Bryansk-Belarus Zone*, which was the second most-polluted zone. It was the area within a 200-kilometer radius from the damaged plant. This zone was contaminated between 28 and 29 April 1986, when rain fell and deposited the radioactive particles, which were being carried by a contaminated cloud that happened to be above the area at that time.
3. The *Kaluga-Tula-Orel Zone*, in the now Russia, which was the region within 500 kilometers from the scene of the accident. The zone was

contaminated by the same cloud, which created the Bryansk-Belarus zone.

The radioactive cloud, although weaker, continued on its journey North, and radiation was now being detected beyond the Soviet Union's borders. Sweden was the first place where, on the 27th of April, 1986, the radiation from Chernobyl was detected by workers who were monitoring a nuclear power station. When the news got out, all nuclear monitoring points across the planet began testing their areas for radioactivity. Other countries where the Chernobyl radiation was detected included: Great Britain, Belgium, the Netherlands, South and Eastern Switzerland, Austria, Scandinavia, North America, Japan, France, Portugal, Spain, and Southern Germany.

The Liquidators

Immediately after the fire had been put out, and plans to evacuate the affected area were under way, the Soviet authorities began the process of discarding the contaminated debris. Priority was given to the plant area where contamination was unimaginable. The roof of the destroyed reactor contained graphite and fuel remains, as well as highly-contaminated debris. It was decided that the best way of dealing with the rooftop debris was to throw it into the remains of the core, and cover it up with neutron absorbers. Additionally, houses that had radioactive particles on the roofs had to be cleaned. Affected trees needed to be felled and destroyed. The streets, too, required cleaning, to minimize the contamination of more people and animals. All the pets in the contaminated area had to be killed and buried to avoid the radiation they had accumulated from contaminating their owners. Owing to the mass demand of mass labor for all these exercises, the Soviet Union called upon willing people to help.

Over 600,000 men and women from across the Soviet Union showed up for the humanitarian cause. These people came to be known as the "liquidators". They would help to limit the short-term and long-term damages, which had resulted from the disaster. On its part, the Soviet Union promised them compensation in terms of honorary medals, priorities in employment, free healthcare, and handsome retirement benefits.

Below is a breakdown of the military and civil liquidators alongside the roles they played in suppressing the after-effects of the Chernobyl Nuclear Power Plant catastrophe:

- **Plant operators**: They showed other emergency responders like firemen around the plant. This not only helped them to assist in evacuating survivors and maintaining the surviving reactors, but also avoiding areas that were likely to be contaminated.

- **Firefighters**: Apart from being amongst the first responders, the firemen kept battling the fire for 10 days until the graphite fire went out. They also hosed the streets and houses with water to wash the contaminants off.
- **The media**: Photographers helped with the documentation of the destruction to serve as a reminder of the fateful events. There were performing artists, who entertained the other liquidators to boost their morale. The media also relayed information from one point to another.
- **Foreign engineers and medical practitioners**: They arrived in Ukraine to help in mediating the effects of the disaster. The engineers worked with nuclear scientists as the medical professionals helped with health-related issues.
- **Soviet Army**: The army personnel were assigned in many areas. Some of the recorded tasks included the deactivation of the

damaged reactor, and the removal of the contaminated debris around the plant, including the roof.

- **Civil and military sanitation personnel**: Female janitors from the sanitation ministries were tasked with collecting and destroying the food left in the abandoned residences to curb the spread of diseases and contamination. Special groups from the military were tasked with the extermination of domesticated animals, which had to be left behind.
- **Civilian workers and engineers**: A group of miners agreed to dig an underground tunnel beneath the destroyed reactor to facilitate the construction of a cooling system. Transportation workers drove buses that evacuated people from Pripyat, and other affected areas. Construction workers would later work for months to create an enclosure for the damaged nuclear plant.

- **Civil aviation and Soviet Air Force units**: They played the important role of providing helicopter operations on the damaged building, monitoring atmospheric contamination, transporting personnel, and delivering the material that was poured to put out the graphite fire.
- **The police and local troops**: They played major roles in evacuating the populations, providing control access, and maintaining security throughout the time of the disaster.
- **Containment**

  When the authorities realized the scope of the disaster, mediation efforts were launched immediately to reduce the consequences. One of the first countermeasures was to start cleaning all the contaminated areas, and dispose of the waste. Members of the Soviet Forces and local troops worked together with the firefighters to spray the building with decontaminants and water. Contaminated vehicles and tools used in the cleanup exercise were taken and buried at hundreds of disposal sites around Chernobyl. This would soon create another problem, since there was too much to bury. In addition, burying radioactive waste risked spreading the contamination into the ground.

- Another area that needed fast mediation was the agricultural sector. Due to the reluctance of the authorities to acknowledge the disaster and implement countermeasures immediately, contaminated fodder was fed to animals. This transferred radiation to the animals, and later found its way into the human body through the consumption of animal products, such as milk. When containment was finally initiated, contaminated milk was rejected at the collection points. Clean fodder was also produced alongside chemicals that trapped caesium, and kept it from spreading.

- Next, restrictions were applied in forests around the contaminated zones. This move was aimed at controlling activities such as harvesting products from the forest, and hunting wild animals. Wild meat had to be inspected before it was allowed for consumption. Complete access restrictions were implemented around the Red Forest and other severely damaged areas.
- One of the most difficult countermeasures that the Soviet authorities tried to implement was protecting water sources from being contaminated. Most of the applied methods failed, since radioactive material leached into the soil and joined water bodies. Additionally, when it rained, there was no way to control the water. Eventually, they settled on switching to sources of uncontaminated water, some of which were pumped from Kiev. Finally, the direct consumption of fish was restricted.

Removal of Contaminated Debris

A few months after the accident, the towns of Chernobyl and Pripyat were deserted, save for the daring liquidators, and a few animals that had survived the extermination. The focus was now turned to clearing the radioactive waste, particularly around Reactor Number Four. The roof was the most dangerous place in Chernobyl, and had over 100 tons of toxic debris. It was necessary to get rid of the debris to curb further spread of contamination, and also allow for the construction of a structure to enclose the reactor. In addition, the Soviet Union preferred decontaminating the areas, rather than waiting for natural decontamination. They hoped to revive the land, and resettle the people who had been evacuated.

Due to the high radiation on the roof, the debris cleanup exercise was initiated by robots. However, every robot that was placed on the roof malfunctioned a few minutes later due to the radiation. Over 60 robots provided by the Soviet Union were tried, but they all failed. The only alternative that remained was to use some of the liquidators to shovel the debris into the reactor. Protective gear was provided to some military personnel, and they began working on the roof. They later came to be known as "bio-robots." To reduce their exposure to the deadly radiation, the liquidators worked for a maximum of 90 seconds, before being replaced. Unfortunately, due to a shortage of volunteers, some of the men worked for longer periods, leading to exhaustion of their lifetime limits of radiation exposure.

After the clearing of the debris, the construction of an enclosure of the reactor, later nicknamed "Sarcophagus", began immediately. The reactor needed to be sealed off, because there was an assumption that, due to the unused materials in the core, the nuclear fission could re-ignite itself, and spew more poisonous gases into the atmosphere. The construction required over 250,000 liquidators, and it took months to complete the Sarcophagus. This was to become the biggest single civil engineering project in the world. It was during this construction that a clip of a helicopter's rotor colliding with the cable of a construction crane and crashing to the ground was recorded. All four crew members perished. As of December 1986, the project was complete.

Other debris and contaminated tanks, cars, trucks, and helicopters used during the cleanup were buried in different sites, including two excavated sites, which would have housed Reactors Number Five and Six. However, some of the vehicles remain parked in the last spots they were left by the liquidators. In 7 months', time, the debris removal exercise was complete, and Chernobyl was deserted.

Chapter 6: Consequences of The Accident

Effects on Human Health

Premature Aging

According to multiple studies conducted in Belarus, Ukraine, and Russia, ionizing radiation has the ability to speed up the process of aging. The ionization affects the cell function and cell structure at both the genetic and molecular levels. The effects of this process cause changes that are similar to those brought by the normal aging process, such as changes in the nerve system, changes in fat metabolism, variations in the working of the immune system, the process of repairing DNA, and the reactions of free radicals.

The studies further showed that a significant number of the people that got exposed to high doses of radiation, particularly the liquidators, died 10 to 15 years earlier than they would have under normal circumstances. Some of the health complications that resulted from the exposure, which led to faster aging and death included:

- Instability of the antioxidant system in the body which repairs the chromosomes of damaged cells
- Faster aging of blood vessels, especially the coronary vessels and blood vessels in the brain
- Hampered higher intellectual cognitive functions due to the damage to the nervous system

Cancer

In the areas that were highly contaminated, such as Belarus and Ukraine, there was substantial evidence that the rate of cancer cases went up. This is according to a national cancer registry that has been in existence in Belarus since 1973. The records showed that, before the Chernobyl nuclear disaster, the rate of new cancer cases per year was about 160 per 100,000 people. In Chernobyl, the rate was approximately 218 cases per 100,000 inhabitants.

However, after 1986, the rates seemingly increased. For instance, Gomel, one of the hardest-hit areas in Chernobyl, saw a 56% increase in bladder, lung, thyroid and colon cancers. Before the accident, only 148 cases were reported for every 100,000 people. After 1986 disaster, the cases shot up to 225.

Infants born during and after the accident were not spared, either. Prior to the explosion, cases of brain tumors in children under 3 years had been recorded as 9 between 1981 and 1985. Alarmingly, from 1986 to 2002, the cases shot up to 179. The rates were even higher in nursing infants.

The other population which proves that cancer rates increased in, is the liquidators. This group was exposed to high doses of radiation for extended periods. Studies conducted on this group, against a control population, showed that the liquidators had higher rates of thyroid, kidney, bladder, lung and colon cancers. Most of the liquidators who lived past the year 2000 were invalids, and a majority of them were suffering from different cancers.

Nervous System Damage

A Belarusian psychiatrist named Minsk Kondrashenko warned that the Chernobyl disaster would affect the nervous systems of some people who got exposed to high levels of radiation. The radiation, he revealed, caused organic damages to the brain. Some of the symptoms that the public had when they first got exposed to radiation included headaches and malfunctioning nerves. However, the Soviet Union's medical fraternity ignored the occurrence, and blamed it on "radiophobia", the fear of radiation sickness.

Decades later, a frightening truth was revealed: Over 48% of post-mortem examinations conducted on liquidators upon death showed that their deaths were caused by problems with the blood circulatory system, such as clots. Surviving liquidators were known to feel dizzy at times, due to lesions present in their nervous systems. In fact, during the liquidation process, most drivers had to stop working, because they dozed off while driving through highly-contaminated areas.

Psychological Complications

In January 1993, *The Moscow Times* revealed the findings of a study conducted in St. Petersburg, which showed that over 80% of liquidators suffered from different psychological complications. About 40% percent of them suffered from neural disorders like memory loss, while the majority suffered from loss of concentration, memory dysfunction, depression, and dysphasia. The possible explanation provided by the radiation professionals is that blood flow to their brains was somehow reduced. Additionally, these psychological disorders were more prevalent in the liquidators than the rest of the population.

A significant number of the survivors of the disaster who developed acute radiation syndrome showed changes in their left cerebrums. This condition led to the emergence of schizophrenia and chronic fatigue syndrome in most of them. Further studies showed that the most affected people were those that got exposed to radiation levels above 0.15 Sieverts and below 0.5 Sieverts. There was a similarity between such patients and survivors of the Bosnian and Gulf Wars, where, in both cases, Uranium-238 was present in the air.

Genetic Mutation

Unusually high incidences of mutations have been noted in the genetic materials of children born by parents who were exposed to between 50 and 200 milliSieverts of radiation. This dose is equal to the exposure that nuclear plant workers accumulate in 10 years. The cases are extremely common in the children of liquidators. In an investigation conducted by Russian scientist Professor Sheban and Prilebslaya, his colleague, the children of liquidators showed more mutation effects as compared to children of unexposed parents. The children of parents who got exposed to the quoted doses displayed higher rates of mental disturbances, leukemia, behavioral deviations, metabolic and endocrinal complications, congenital deformations, and cancer.

Infant Mortality Rates

In the decade leading up to the Chernobyl accident, mortality rates in Europe had been on the decrease. The improvement was attributed to improving living conditions and better medical care. However, after Chernobyl exploded, the affected areas in Ukraine,

Belarus and other parts of Europe showed an increase in infant mortality. One year after the disaster, the number of perinatal deaths and stillbirths around Chernobyl, Ukraine, and Belarus increased. This was blamed on the exposure of expectant women to caesium and strontium. There were further differences in mortality rates between the areas that were highly contaminated, and those that received lower concentrations of radiation.

Further away in Germany, where some of the radiation was detected, a similar effect was discovered. In 1985, the infant mortality rate in Berlin was 10.6 infants per 1,000 births. In 1986, the number shot up to 12.5. In the same country, there were more stillbirths in the South, particularly Bavaria, as compared to the North, which had received less radiation. In addition to the poisonous cloud that had been sweeping over Europe, this occurrence was blamed on farmers who had fed their animals with contaminated food.

Thyroid Diseases

Thyroid diseases were some of the delayed health consequences of the nuclear accident, but they later came to be quite prevalent. Complications with the thyroid gland began showing up in Belarus in 1990. Prior to the accident, only two cases of thyroid cancer would be reported every year in the country. Three years after the accident, the number rose to 7. One year later, the rate skyrocketed to 22, and that was when the focus was shifted to thyroid complications arising from the meltdown. In addition to thyroid cancer, other effects such as: Hypothyroidism, autoimmune thyroiditis, and swelling of the thyroid gland were on the rise.

Upon the revelation, the World Health Organization (WHO) conducted studies in some of the highly-contaminated zones. In its findings, the WHO revealed that there were steep increases in thyroid diseases, especially in children living in those areas. In addition, the thyroid complications were abnormally aggressive, and would affect other parts of the body. A similar aggressiveness of thyroid complications was seen in adults. In 1980, in Belarus, only 1.2 adults out of every 1,000 reported developing thyroid cancer every year. In 1990, the number rose to 1.96. By the year 2000, the annual rate had risen to 5.67.

Death Toll

On the first day of the accident, only two fatalities had been reported. The two were plant workers who had been directly killed by the explosion and the resulting heat, and not due to radiation. The deaths caused by radiation began being reported in May 1986, when the thousands of liquidators from all over the Soviet Union began arriving. Close to 300 people, most of whom were firemen, were hospitalized within 1 day of the accident. Most of them had been first responders, and due to their unpreparedness, they had approached the nuclear disaster zone as a normal electrical fire. This exposed them to extreme levels of radiation, estimated to be as high as 10,000 times the recommended levels. Some of them were so highly contaminated that they became sources of radiation and poisoned the rescue and medical personnel who tried to save them.

The first measure that was used to classify the affected people was one's distance from the epicenter of the blast. The Soviet medical personnel would rate people that had been within 1.6 kilometers of the Chernobyl disaster zone to have acquired a lethal dosage of gamma rays. Such people were expected to die, and most did not live beyond two months. The plant workers that survived, as well as the firefighters and police officers who rushed to the scene after the explosions, were in this category.

The second category was people who had been within 5 to 6 kilometers from the site. This group was considered to have acquired medium doses of beta rays. People from Pripyat were mostly categorized in this group. They had at least 50% chances of surviving, but would not be likely to survive for more than five years, due to intestinal damage and bone-marrow destruction. The third category was for people who had been within 8 to 12 kilometers. Depending on the amount and duration of exposure, they would either suffer mild symptoms and recover, or develop serious complications like cancer later in life.

Radiation affects the body by burning the skin and the underlying organs. In severe cases, it reaches deep into the bone marrow and destroys it. The bone marrow helps the body to strengthen the immune system, clot blood, and manufacture the red blood cells. When it is destroyed by radiation, it can lead to death in a few weeks. The firefighters who had approached the fire without protective gear had their whole bodies, including their internal organs destroyed. Their lungs and intestines would rot, leading to starvation and suffocation.

In estimation, immediately after the explosions, 300 people acquired acute radiation poisoning. 29 of them died, as the liquidators were streaming into Chernobyl and Pripyat. About 70 plant workers who were within the Chernobyl plant suffered gamma and beta ray exposure, and most did not live beyond 10 years. During the liquidation process, it is believed that thousands of the volunteers were knowingly or unknowingly exposed to harmful levels of radiation, and over 4,000 did not live past 15 years. Thousands of others were said to have become disabled, and they live in agony up to this day.

The actual death toll of the Chernobyl disaster remains a controversial issue. While most credible bodies like the WHO place the official toll at 31, other sources argue that the toll was much higher, having been covered up by the Soviet authorities. There are more speculations which state that the death toll of the short-term deaths, plus those caused by complications arising from radiation, run into the thousands. To this day, the actual number of fatalities remains unofficial.

Effects on the Environment

Deposition of Radioactive Material

During the 10 days that the graphite fire had been burning, poisonous isotopes such as plutonium, strontium-90, caesium-137, and iodine-131 were constantly being released into the atmosphere. The most severe contamination affected approximately 200,000 square kilometers of Europe. Ukraine, Russia, and Belarus were the most affected countries. In the areas where it was raining during the 10 days, there were more radioactive deposits. Again, since plutonium and strontium particles are heavy, they were distributed within 100 kilometers of the plant. As such, the distribution of the radioactive material was uneven.

The risk that radioactive material poses to the environment is determined by its half-life. This is the total time taken for half of the present radioactive material to decay. The materials that had short half-lives decayed, but others with longer decay times exist and will continue existing for tens, hundreds or thousands of years. For instance, plutonium, as well as its decay products, will remain dangerous for hundreds or thousands of years later, followed by caesium-137 and strontium-90. Until then, some of the contaminated areas will remain dangerous for humans, animals, and plants.

Urban Life

As radioactive material rained down on the areas around Chernobyl and Pripyat, it collected on roofs, roads, parks, pools, lawns and all other available infrastructure. The contamination would have been dangerous to people, and so they had to be abandoned. In the end, all the houses in the danger zone were abandoned. The shopping centers and recreational facilities were left empty. Fields and parks were left unattended, and are now overgrown by trees and other plants.

There were several attempts made to reduce the contamination, such as hosing the houses, and scrapping the topsoil. However, these were not enough, and the urban areas remained uninhabitable. As time went by, the radiation levels on the infrastructure reduced due to further decay of the material, continuous cleaning, and weather effects of rain and wind. A new problem emerged because the material dropped from the buildings and other facilities, but ended up on the ground or in water and sanitation systems, causing secondary contamination. As such, some of the urban areas remained deserted, although a few of them have been repopulated after radiation levels dropped to acceptable levels.

Forests

Before the accident, about 40% of the Chernobyl area around the plant comprised of a thick forest cover. Over 80% of the forest consisted of spruce and pine trees. Most of the trees were approximately aged between 30 and 40 years. When the radiation leak began, the forest filtered significant amounts of the material from the atmosphere. Due to the nature of radioactive material, most of the trees died, especially the coniferous species, since they are more sensitive to radiation. In addition, the accident occurred in spring when the coniferous trees exhibited more radiosensitivity, as opposed to winter and autumn, when they are usually dormant. Within a few days, about 4 square kilometers of trees died, giving rise to the name, "The Red Forest."

Herbs

Unlike forests that have tall trees that acquired enough radiation to kill them, herbaceous vegetation around the Chernobyl Nuclear Power Plant appeared to withstand the disaster at first. This was partly because the trees filtered most of the radioactive material before it got to them, or because their foliage generation was higher than that of trees. In short, they could replace damaged foliage faster. However, after a few weeks, the herbs began dying out. It was discovered that the radioactive particles had saturated the top layer of the soil, and affected the food and water intake of the plants. In the highly-contaminated zones, new herbs would not grow past the shooting stage, and they would wither and die.

All plants in the contaminated area will remain dangerous for human and animal consumption mainly due to the presence of cesium-137, which decays very slowly. This element is readily taken up by plants, and when the plant products are consumed by animals, birds, and humans, contamination occurs. In Finland, Sweden, Russia, and Norway, deer meat was contaminated after the accident, when the animals consumed plants with radioactive content.

Aquatic Life

Water bodies received equal or more amounts of radioactive materials as other areas, despite showing lower concentration levels. The reservoirs, lakes, and rivers around the plant showed a faster reduction of contamination levels than the land. This phenomenon is attributed to the materials being diluted by the water, faster decay due to water, absorption by the surrounding soil and rocks, as well as vast distribution by the moving water.

Aquatic animals like fish absorbed a lot of iodine, but due to its fast decay, the levels declined rapidly. On the other hand, radioactive caesium accumulated in them, leading to high concentrations of radiation in fish and other organisms in the food chain, even in distant areas such as Germany and Scandinavia. All the same, the significant level of strontium-90 in the fish did not affect humans, since the element is deposited in the bones.

The aquatic bodies in the contaminated zones remain highly dangerous since they are constantly replenished with strontium-90 and caesium-137, which come from soils and floodwater from the abandoned regions. As time goes by, the levels of these elements decrease due to dilution and decay, especially in large water bodies such as the Baltic and Black seas.

Insects

Invertebrates are known to be highly resistant to radiosensitivity, as compared to all other organisms. However, due to the high concentration of radionuclides on the top layer of soil after the accident, insects that live on or near the soil were largely affected. The insects that live under the soil were kept safe by its shielding effect, which kept the radiation from penetrating under the surface. Reports indicated a thirty-fold reduction of sexually-immature insects and mites that lived on the soil in the Red Forest, as compared to the forests that were over 30 kilometers away from Chernobyl. The number of nymphs and larvae in the soil around the site reduced sharply as well.

In terms of mutation and biological dynamics, insignificant changes were observed in insects. However, aphids appeared to be affected. In a field of birch trees that was home to about fourteen species of the insects prior to the catastrophe, only two species remained by the year 1994. The reduction of the insects was attributed to radiation as well as loss of habitat and food, as trees and herb numbers decreased.

Birds

Significant pathological effects of radiation on bird species living around the Chernobyl disaster area were noticed. For example, an increased concentration of iodine and strontium in the eggshells of pigeons and crows deformed the eggs. There was also an increase in partial albinism in nesting and adult barn swallows, as compared to those that lived far from the contaminated area. This condition was rare before the accident. In addition, the population of barn swallows had gone down, although no direct effect of radiation was found to be the cause. Secondary factors such as changes in habitat and community structure were listed as the likely causes.

Mammals

Populations of small mammals like rodents reduced sharply immediately after the accident. This affected the small animals that lived near human populations or depended on forests and shrubs for habitats and food. At first, high radiation doses caused massive deaths, followed by the lack of food as plant life declined. Surprisingly, though, the small mammal populations began springing back in 1987. The evacuation of human beings, migration of animals from uncontaminated areas, and the regeneration of plant life contributed to the repopulations.

Significant biochemical changes affected most of the mammals living in the contaminated area. Liver cirrhosis cases in animals increased from 1987. All the horses and cattle that were left behind in the high-contamination areas as their owners were evacuated, died from thyroid-related complications. Those that received small doses of radiation portrayed sign of stunted growth, deformation at birth, and hypothyroidism. Some animals were so severely affected by abnormalities that they died a few hours or minutes after birth. In addition to cattle and horses, roe deer and wild boar were also affected by similar issues.

Evacuation

Soviet authorities initially treated the accident like a local emergency that only affected the Chernobyl power plant. A few hours into the incident, even after nuclear scientists were aware of the radioactivity in the atmosphere, they were reluctant in informing the plant workers and their families to take precautions or move from the area. Evacuation plans were hastily hatched 36 hours after it became apparent that the burning reactor would not stop releasing radioactive material into the air.

The first perimeter of those to be evacuated was initially 5 kilometers. No quantifiable reason for the evacuation was in place before this decision. It is said that the radius was decided just because the area was in close vicinity of the damaged plant. However, some hours into the process, when radiation measurements were taken, the evacuation radius was extended to 10 kilometers, and then to 30 kilometers. This 30-kilometer radius would later come to be known as the "exclusion zone".

The following Sunday after the explosion, 1,100 buses, mostly borrowed from Kiev, arrived in Pripyat to evacuate the inhabitants. The buses formed a 19-kilometer line, as the 50,000 residents boarded them in an exercise that took only 3 hours. The exercise was hurriedly commanded, as the officials hoped a solution for the radiation would be found in a few days' time. Therefore, the residents were expected to pack a few belongings and get into the buses as fast as they could, leaving most of their precious belongings behind.

Contrary to what the officials assumed of the situation, it got worse rather than improve with time. As the graphite fire raged on and radiation levels proved to be high in areas much further from the smoldering reactor, they were forced to expand the evacuation radius to the town of Chernobyl, and parts of Ukraine and Belarus. By the end of May 1986, over 350,000 people had been evacuated from all the contaminated regions to Zhytomyr in Ukraine, and some provinces around Kiev.

After completion of the evacuation process, controversies began tainting what had previously appeared to be an impressive achievement. First, the residents complained that the authorities had taken too long before revealing the truth. This had resulted in a significant number of people getting sick from the radiation. Others were unhappy with the "lie" that they would only be away from their homes for a few days, only for it to become permanent. As more revelations were made, family members spoke of how they were separated from their families as the officials hurriedly loaded them into buses and trucks. Additional reports indicated that some people were even left behind because they could not abandon their pets and animals which had been left behind.

Economic Effects

The Chernobyl meltdown, the countermeasures applied, and the policies that were adopted in coping with the situation, caused serious implications on the

Soviet Union and the three successor nations: Belarus, Ukraine and Russia. In as much as the biggest impact was felt in this region, more countries beyond the Soviet Union also suffered financial losses due to the radiation. In estimation, the disaster cost the Union well into the hundreds of billions in sustained economic losses.

Agriculture

Radiation in the air, soil and water bodies meant that agriculture in the affected zone had to be stopped. In total, an estimated 784,000 hectares of productive agricultural land was rendered unusable after the April 1986 accident. Food shortages hit the Soviet Union. Timber, which was a major economic product in the region, could no longer be harvested from the region, and 690,000 hectares of forest were wasted. While the authorities successfully remediated some of the fields, and the production of safe food was restored, the costs of the required chemicals made farming unsustainable. In addition, the stigma that came to be associated with the Soviet Union meant marketing their produce would be difficult. In the long run, farming weakened, jobs were lost, production slowed down, and revenue fell.

Resettlement

Resettlement costs began the moment the first evacuation was done. It was up to the government to provide means of transport for the people and their belongings. After evacuating them, the authorities had to find shelter for them. Before a few towns were constructed to settle the evacuated people, the governments had to pay for the houses from its own coffers during a time of disaster. In 1991, the rehabilitation programs and other related projects stalled as they proved to be unsustainable. This was blamed on the growth in the number of victims to seven million. They were allowed health benefits, special allowances, and pensions—a combination that eventually overwhelmed the Soviet Union.

Social Welfare

A number of people remained in the mildly-contaminated areas of Chernobyl, and the authorities promised to support them. Due to the contamination of the ground and water, the USSR authorities decided to limit the consumption of the food they cultivated. They improvised measures such as free meals for students at school. There were also cash benefits to help the people in acquiring safe food, since most of what they produced was not entirely safe. Finally, the government began sending free supplies of fruits, oil, dairy products, and meat to them.

Direct Costs

The urgency of the disaster meant that the affected countries needed to act fast. This resulted in impulse spending to counter the effects of the life-threatening situation. In the case of the Soviet Union, some of their direct costs include:

- Providing transport and communication for the liquidators and the evacuated citizens

- Compensating liquidators who came to help in the rescue efforts
- Purchase of equipment to lift materials to the roof of the burning reactor
- Building a sarcophagus for the damaged reactor to seal the radioactive materials inside
- Resettling people and constructing new towns for them
- Funding research to identify and eradicate the contamination
- Sealing the exclusion zone to keep contaminated animals and materials from getting out, or people from going into the zone

Psychosocial Impacts

Trauma

Evacuation might have saved 500,000 people from the effects of radiation, but it also created a new problem. Most of the resettled people suffered from differing traumas. In as much as free accommodation was provided, as well as compensation and hefty promises, they felt that the entire process was not justified. For example, they felt that they should have been allowed to choose their places of resettlement. Loss of employment meant that the once-independent people had to start looking for alternative sources of livelihood, or look upon the government to provide for them. In addition, it was hard for the elderly resettlers to adjust to the new environments.

Distorted Demographics

Another problem caused by the resettlements is that the natural order of demographics was distracted. When the masses were moved into foreign regions, the younger generations who were entrepreneurial, educated and skilled left their new homes in search of better livelihoods and opportunities. They left the older generations behind to cater for themselves. As such, in the new settlements, the populations of senior citizens were very high. Due to limitations caused by age, the elderly people left behind could not partake in economic activities to eliminate poverty and resuscitate the economy. In short, the areas remained underdeveloped and dormant.

Psychological Distress

The accident at Chernobyl had serious mental impacts on the public, and it changed the way people perceived life. For instance, since the abrupt disruption of their daily routines, the people lost the sense of being in control of their lives. They felt like someone else had moved in to control them. In yet another instance, due to their exposure to radiation, the population lived with the notion that their life expectancies were short. They lived with the fear that some radiation-related illnesses would affect them any time, and destroy their lives. Finally, since the people were moved to foreign lands, some of them faced hostility in their new homes. This made them feel unwanted, and they also lost their sense of belonging. Even in areas where the reception was favorable, it took them a lot of time to get used to the environment. All these factors added to their psychological distress.

Fear

After word got out that food and water in the contaminated areas might have contained radioactive material, anxiety gripped the public. People would not purchase meat that did not carry a stamp from the USSR Health Ministry. Shoppers were reported to be curious about the source of food from which they purchased. In some instances, they demanded to see the passports of the vendors, to be sure that the food originated from beyond Chernobyl. The consumption of meat and milk reduced sharply, as far as Leningrad. Although there was plenty of meat in the 1986 summer, people were hesitant to consume it. In 1987, there was widespread renewed fear over contamination due to the spring floods as the snow melted. There were rumors of people stocking up large quantities of bottled water in anticipation of the foreshadowed disaster.

Chapter 6: Investigations and Trial

The Investigation

As the confusion that followed the accident continued, the Soviet Union leaders launched investigations almost immediately. They appointed Boris Shcherbina as the head of the inquiry to investigate the causes of the accident, and present a report to them. He was the deputy chairperson of the Council of Ministers, and the chairman of the Bureau for Fuel and Energy. However, he would not be the main figure to receive credit for the investigations, because of one Valery Legasov, a Soviet nuclear physicist and chemist. Legasov came into the limelight due to his concerns over the safety of the public living around the plant. He was also very vocal in providing solutions for putting out the radiation fire in the destroyed reactor.

It is not clear how he came to take over the investigations, but his stubbornness in calling for transparency between the public and the investigating commission played a key role. Legasov was the deputy director at the Kurchatov Institute of Atomic Energy. After receiving the news about the disaster, he initiated his own investigations, which saw him appointed to the special commission that would be investigating the disaster. He would later become overly influential in providing solutions, as well as highlighting the potential causes of the explosions.

The first thing that Legasov noted about the disaster management approach used is that it was highly disorganized. There was insufficient equipment such as radiation detectors and respirators. Delays occurred before the evacuation orders were made, and radiation doses in people were increasing. More importantly, the government was reluctant in accepting the reality of the situation, and tried to downplay the situation. Generally, the management of the crisis was bad and lacked organization.

Legasov's Report

In a few days' time, Legasov presented a report to the government. In his investigations, he had come up with two conclusions:

First, his team had found that the design of the RBMK-1000 reactor had serious faults, which might have led to the disaster. The reactor type was known for its instability, and had even been banned from operations everywhere else but the Soviet Union. The report further revealed that nuclear scientists had warned the Soviet government against using the reactor, but the warnings had been ignored.

The second potential cause of the accident was blamed on the semi-skilled workers who had been assigned to perform the safety test. On the day of the test, the team that had been assigned to handle the task committed serious mistakes, which reduced the safety margin of the already-faulty reactor. The mistakes had pushed the safety margins to their limits, and the explosions occurred. However, most of the blame was placed on the deputy chief engineer, Anatoly Dyatlov, for conducting a test without authorization, and overlooking the recommended safety procedures.

He would later present the same findings in August of 1986, in Vienna, during a conference hosted by the International Atomic Energy Agency. Legasov stood by the fact that the faulty design, combined with human error, collectively caused the accident. His transparency was applauded by the international community. However, back at home, more efforts were being implemented to downplay the magnitude of the disaster. Legasov also received serious backlash from power players in the Soviet Union for his honesty.

The Trial

Following the report, six people were to stand trial for their roles in the disaster. They were:

1. Victor Bruchanov (52 years old): He was the director of the Chernobyl Nuclear Power Plant
2. Nikolai Fomin (50 years old): He was the chief engineer at the Chernobyl Nuclear Power Plant
3. Anatoly Dyatlov (56 years old): He was the deputy chief engineer at the Chernobyl Nuclear Power Plant
4. Aleksandr Kovalenko (45 years old): He was the chief engineer at the reactor hall number 2
5. Yuri Laushkin: He was an inspector at the plant
6. Boris Rogozhkin (53 years old): He was the station shift supervisor

A public trial began on July 7, 1987. It was held in a makeshift courtroom inside the Chernobyl exclusion zone, about 11 kilometers from the plant. Some of the participants were a panel of judges, two assessors, a government lawyer, and eleven nuclear experts. Journalists were allowed in only during the first session and the last one, which was held on July 29 of the same year.

Accusations

The six respondents were accused of approving and executing a risky scientific experiment without proper authorization and supervision, leading to the disaster. This caused the reactor to explode and pollute the environment. As a result, the plant was destroyed, hundreds of thousands of people were evacuated, thirty-one people died, and thousands of others got sick.

The second accusation was that, once the accident occurred, the six accused ignored the appropriate countermeasures for the plant workers and the people living near the plant. Rescue operations were delayed, and workers were allowed to stay in the contaminated area. Additionally, they relayed false information in a bid to conceal the actual magnitude of the disaster. For instance, they provided lower-than-recorded levels of radiation in the Chernobyl area.

The third accusation was that, despite numerous accidents occurring in the Chernobyl Nuclear Power

Plant before 1986, they were not recorded nor analyzed. Worse still, the management of the plant denied the staff necessary professional training, and ignored work ethics.

Verdict

On the final day of the trial, the presiding judge, Raimond Brize, had a 90-minute verdict ready. The summary of the convictions was as follows:

- Viktor Bruchanov was sentenced to 10 years in jail for criminal negligence and violation of safety rules. He received an additional 5 years for abuse of power.
- Nikolai Fomin and Anatoly Dyatlov received 10-year sentences for unauthorized experiments and violation of safety regulations.
- Boris Rogozhkin was handed a 5-year sentence in a corrective labor camp.

- Alex Kovalenko was sentenced to a corrective labor camp for 3 years.
- Yuri Laushkin would also serve in the same camp for 2 years.

On the same day, the spokesperson for the Chernobyl power plant said that three more trials would be opened to explore additional issues such as reactor construction and design, plant security failure, and poor post-disaster management, such as the delayed evacuation of the public.

Beyond the courtroom, the Communist party expelled 27 people, and 67 others were reprimanded for the roles they played in the April 1986 Chernobyl Nuclear Power Plant disaster.

Legasov's Death

On April 28, 1988, a day before Valery Legasov was to present his findings on the design flaws of the RBMK-1000 reactor, he was found hanging from a rope inside his house. This was two days after the second anniversary of the catastrophe. Some of the possible causes of his suicide include backlash from authorities in the Soviet government, continued use of the faulty reactor despite his campaign against it, and failure of the government to implement recommended international nuclear safety regulations.

Upon his death, he was posthumously awarded the "Hero of the Russian Federation" award in 1996 by Boris Yeltsin, the President of Russia, for his courage in countering the disaster, and restoring sanity in the nuclear energy industry.

Chapter 7: Causes Of the Accident and Rectifications

Following the death of Legasov, the Soviet government accorded the matter of Chernobyl the criticality that it deserved. Two reports were published to explain the causes of the accident. Upon finding the cause, it would be possible to make necessary amendments so as to prevent future accidents. The first report, which was later dismissed, placed the blame on the plant operators who were on duty during the test. It had multiple flaws, and so it was never accepted. In 1991, a commission was formed by the USSR State Committee to reassess the circumstances and causes of the accident. One year later, a second report was released. This time, it addressed the design failures which Legasov had always highlighted, and some critical human errors as the causes of the meltdown.

Below is a discussion of both causes.

Design Flaws

Positive Void Coefficient

The biggest flaw in the design of the RBMK-1000 reactor was its extreme positive void coefficient. The void coefficient measures the response of a reactor to the increase in steam formation inside the coolant (water). Other reactors are made to have a negative void coefficient, such that when steam bubbles form in the cooling water, the reactivity of the fission process slows down. This is because steam is not a good absorber of neutrons like water, and so, fewer uranium atoms will be split, resulting in lower power output.

The RBMK-1000 reactor used graphite blocks to moderate the neutrons by slowing them down. The water in the reactor acts as the absorber of harmful neutrons, and also as the coolant. Therefore, the neutrons would still be slowed down even when there were steam bubbles in the water. As such, since steam absorbs fewer neutrons than water when more of it is produced, the neutrons are not slowed down, and they split more uranium atoms, producing more heat. This would result in higher reactor power output. As a result of the positive void coefficient, the reactor was prone to becoming unstable when power levels were low.

Control Rod Design

The second flaw, as highlighted in the report, pointed at the design of the boron control rods, which were used in slowing down the nuclear fission process. In the reactor's design, the rods had graphite tips, and this made them about 1.3 meters shorter than necessary. The space beneath the tips and the active boron were hollow and full of water. The boron part of the rod was the one responsible for slowing the reaction down, since boron readily absorbs neutrons. Due to this design, when the boron rods were inserted into the reactor from the top, the graphite tips would displace the water and result in the absorption of very few neutrons. As such, the immediate reaction after the control rods were inserted was a sudden surge in reactivity (thus higher power output).

In addition to the design of the control rods, the time that it took before the 7-meter length of the rods was fully into the reactor reduced their effectiveness. Typically, it took approximately 18 seconds for the rod to be fully immersed. During the emergency, the 18 seconds proved too long, and the explosion occurred before the rods could control the nuclear fission reaction.

Containment

The final flaw of the reactor is that, due to its design, the protruding chimney and cranes used in lowering the fuel rods made the construction of containment around it difficult. It was possible, but it would incur more costs and time. As such, containment was ignored. This was one of the international nuclear reactor requirements which the USSR ignored.

While this omission had no direct effect on the cause of the explosions, it would have reduced or prevented the leakage of radioactive material into the atmosphere. The contaminated debris and the unused materials which were spewed around the Chernobyl surrounds would have been contained by such a structure. Unfortunately, there was no way to control the poisonous omission, and the on-site accident blew out of proportion to become an international disaster.

Human Error

Semi-Skilled and Uninformed Operators

It was revealed that the party-in-charge of the design and installation of the RBMK-1000 reactor did not inform the chief engineers and the operator staff of the two design flaws. In their minds, the operators believed that the reactor was perfect, and that the chances of an accident were insignificant.

Low Power Operation

When the reactor operated on low power, it not only became unstable, but also produced xenon, a by-product of the nuclear fission reaction. Xenon poisoned the core of the reactor, and reduced its reactivity. Since the operators were not aware of this, they left the reactor running on low power since the afternoon of April 25.

When it was time to run the test, the xenon poisoning in the reactor caused the lowering of power to incredibly low levels. The automatic power decrease prior to the test was a result of the poisoning.

Blocking the Automatic Control System

The operators shut down the automatic control system, which would have regulated the flow of coolant and insertion or removal of the control rods in the event that reactivity increased or decreased past safe operating margins. By switching them off, the reactor became a manual system that relied fully on human operation.

Due to this blunder, when xenon levels escalated, the reactor could not automatically increase the power so it could burn away. Similarly, when the coolant pressure was too low, the automatic system could not increase it, leading to overheating in the core. In addition, this system would have lowered the control rods in time to curb the increase in output power.

Turning the Warning System Off

Just as the automatic control system was disabled, the reactor's alarm system was switched off. This was aimed at allowing the operators to perform the test without constant warnings of system malfunctions, since they would be deliberate.

This turned out to be a serious mistake, since more malfunctions than those anticipated had occurred. For example, steam pressure was not supposed to reduce at any point in the test. Similarly, the output power should not have been reduced below 700 megawatts. With the warning system turned off, the operators had no way of identifying these additional risks.

Turning the ECCS Off

The Emergency Core Cooling System, ECCS, was turned off, since the operators thought that the turbogenerators would provide enough power to run the water pumps and cool the reactor. It was unnecessary, since the ECCS would only turn on when the core temperatures exceeded safe levels.

When the lack of coolant and increased reactivity caused extreme heating in the reactor core, the ECCS could not mediate it, since it had been disabled. As a result, the heat kept surging until there was enough to vaporize all the coolant.

Connecting All the Coolant Pumps to the Reactor

When all four water pumps were connected to the reactor, it meant that none was independent, and in the event the test failed, there would be low coolant pressure. Instead, one or two of the pumps should have remained connected to the external power supply so that it would provide coolant, in case the pump under test failed.

Removing All the Control Rods

The unexplained loss of power in the reactor made the operators remove almost all the control rods, in a bid to restore the optimum power needed for the test. Unknown to them, the core had become poisoned by xenon, and the power loss had nothing to do with the control rods.

With all the manually-controlled rods out of the reactor, when the xenon began burning away and power began rising, there was no way of controlling the reaction. As such, power doubled every second, and then every millisecond, as the fission process accelerated. Eventually, when they began dropping the rods into the reactor, it was too late, as the power had reached critical levels.

Initiating the Test at Low Power

Dyatlov had become impatient with the test, which had been postponed from the previous afternoon.

When power sunk to 30 megawatts, and later escalated to 200 megawatts, he ordered the test to begin. In so doing, he had violated one of the parameters set for the test that required a minimal operating power of 700 megawatts.

Conducting the test at low power had several implications: First, xenon would continue accumulating, and risk choking the reactor. Second, the reactor was unstable at low power. Finally, due to the low output, the pumps, which were now being run by the reactor, received insufficient power. As such, there was insufficient water in the coolant system.

Pressing the AZ-5 Button

A SCRAM system was part of the reactor design. It was intended for emergency purposes, particularly when the reactor needed instant control of reactivity. The system would be initiated by the pressing of the

AZ-5 button, which would lower all the control rods into the core of the reactor to reduce or stop reactivity.

Due to the lack of information on the design flaws, and instability of the reactor during low power operation, Toptunov pushed the button which began the insertion of all the control rods at once. As the graphite tips displaced the water inside the channels, neutron absorption decreased instantly, and there was a sharp surge in nuclear fission. Inside the coolant pipes, much of the water had turned into steam. The sudden increase in reactivity caused all the water to evaporate, and the resulting high pressure ruptured the pressure channels. More steam and the absence of coolant increased the power by 100 times, and the core of the reactor exploded.

Rectifications of the RBMK-1000 Reactor

The second report that was presented to the USSR State Committee quantified Legasov's demands that the RBMK reactor required some changes in the design. Sadly, the changes were adopted after he was dead. All the remaining RBMK-1000 reactors were soon updated with safety improvements to prevent a second Chernobyl from happening. Some of the improvements included:

The Control Rods

The design of the rods was prioritized in the improvements of the remaining reactors. One of the main problems with the original design of the rods was that they had some graphite tips at the end, which would displace water that would settle at the bottom of the channel where the rods would come to rest when fully inserted. The problem was that, as the rods sunk in, they would displace the neutron-absorbing water, and cause an increase in reactivity, before the rods eventually reduced. In the new design, the rods were made such that no water remained in the channels when the rods were fully retracted. In this way, there would be no imbalance in power during rod insertion.

The Positive Void Coefficient

Measures were also put in place to reduce the positive void coefficient. To achieve this, the designers added 80 to 90 fixed absorbers in the core to help in inhibiting operations during low power. The second measure was to increase the number of control rods required for the operating reactivity margin, ORM from 26 to 30 rods to 43 to 48 in steady operational mode. Third, they increased the enrichment of the uranium fuel from 2 to 4% to sustain the burn-up of fuel even as neutron absorption increased.

Reactor Shutdown Time

The time it took to shut the reactor down, especially during emergencies, was reduced. To do this, the insertion time of the control rods was reduced from 18 seconds to 12 seconds. Secondly, a fast-acting protection system for emergencies was installed. This system, known as the "FAEP", had 24 control rods reserved for emergencies. They responded faster, and would be fully immersed in just 2.5 seconds. Finally, the lack of residue water at the bottom of the control rod channels also helped in improving the time required in shutting the reactor down.

In addition to these three main corrections, some other minor improvements that were done included:

- All the fuel channels in the remaining units were replaced

- The ECCS systems were improved so they became automated, and responded much faster
- All the group distribution headers were replaced, and check valves were added to improve their safety
- The SKALA process computers were all replaced with upgraded computer systems
- The over-pressure protection systems in the cavity of the core were improved, so they detected pressure faster, and let it out in case of excess pressure
- Access to emergency safety systems was reserved only for the plant chief engineers

It is not clear why the lack of containment was not addressed. The remaining reactors at the plant operated without the radiation control covers until their decommissioning in the early 2000s.

Chapter 8: Modern-Day Chernobyl

Closure of the Plant

The Chernobyl Nuclear Power Plant resumed operations after the debris on Reactor Number Four's roof had been cleared of radioactive material. There were local, national and international concerns about using the faulty RBMK-1000 reactors. However, after the USSR implemented the corrections to the reactors, the outcry decreased. Unit Four was entombed in a concrete Sarcophagus. The tomb had 16 tons of plutonium and uranium, 30 tons of contaminated dust, and 200 tons of corium. It was impossible to repair it.

Reactor Number Two suffered extensive damage in the turbine room, caused by a fire in 1991. The Ukrainian government decided to shut it down for good.

The two other reactors remained in operation, and provided the Soviet Union with power for 14 years after the accident. Reactor Number One was shut down in 1996 after an international outcry re-emerged when thyroid cancer and effects of radioactive poisoning from the 1986 accident began showing up. Continuous pressure from international governments led to its closure. After the accidents affecting Units Two and Four, western countries pleaded with Ukraine to close the remaining Reactor Number Three. However, Ukraine was reluctant in shutting it down, since its economy was not stable after the separation from the Soviet Union. Western and European countries went into negotiations with Ukraine, and they promised to fund the country so it could find alternative sources of power. The agreement was signed, and the last reactor was shut down in 2000.

The decommissioning of the three reactors began in 2015. This process involved the safe storage of used fuel, the decontamination of liquid and solid waste generated during the demolition process, and the safe containment of the entire plant in disassembled form until the radioactive material decays fully.

The Chernobyl New Safe Containment

In 1996, the Sarcophagus which had housed the damaged Reactor Number Four since 1986 appeared to have been highly deteriorated. It had undergone multiple repairs over the years until it could not be repaired any more. The fear of radiation leaks from the active fuel and contaminated material called for the construction of a new tomb.

The last-ever repairs for the Sarcophagus were done between 2004 and 2008, as plans for the new confinement were being laid. The new sarcophagus would come to be known as "Chernobyl's New Safe Confinement". This structure would be a steel, arch-shaped building, with a height of 92 meters. The distance from the lower and upper chords would be 12 meters. On the inside, it would 245 meters long, and 270 on the outside. Its width would be about 150 meters, made of vertical walls standing independently from the old sarcophagus. The vast size would not only cover the old tomb, but also allow for the movement of workers and machines as they dismantled the concrete housing inside. Dry air would constantly be circulated in-between the outer and inner roofs, to curb condensation, which might cause corrosion.

Unlike the old confinement, which was hurriedly constructed, the new confinement has specific parameters intended for nuclear entombment. Some of these features are:

- Allow the decommissioning of the old plant safely without affecting the external environment
- Prevent further corrosion and weathering of the old concrete shelter

- Allow for safe disposal and containment of contaminated materials in case the old building collapses
- Allow for the use of remote-controlled machines in the demolition of the contaminated building, and,
- Confine the damaged reactor for over 100 years to allow the radioactive materials to decay

Construction of the new shelter began in September 2010, at the cost of approximately $1.7 billion. The major assembly was done beside the old shelter, and upon completion, it was slid using rails to its final resting position. In July 2019, Chernobyl's new home, weighing over 36,000 tons, and which is the largest mobile structure ever built on land, was handed over to Ukraine. The project was collectively funded by the European Commission, as well as received funding and contributions from 45 nations across the world.

Tourism

Since the closure of the last reactor, people began visiting the exclusion zone, which was under the care of the military. With time, people began flocking in the thousands every year, visiting the town of Chernobyl and Pripyat to experience first-hand, the scenes of hurried evacuation as people fled from radiation. Chernobyl had become a dark tourism destination. "Dark tourism" refers to the act of touring popular sites that are associated with death, such as the 9/11 Memorial in New York, the Genocide Memorial in Rwanda, and the Nazi Concentration Camps in Germany.

The exclusion zone, which opened doors for visitors in 2011, is the perfect reminder to the visitors of the potential of nuclear energy, which is constantly being explored more and more every day. In addition to the deserted towns and recreational facilities, visitors frequent the different memorials set up in the towns to honor heroes who sacrificed their lives to save others and rescue the town at its hour of need. One of the most popular memorials is one dedicated to the firemen who were the first responders. They would be the first victims of radiation poisoning.

In July 2019, during the inauguration of the Chernobyl's New Safe Containment, the Ukrainian President, Volodymyr Zelenskiy, promised to make the exclusion zone, Chernobyl, and Pripyat official tourist zones by making them more tourist-friendly. 2019 saw an influx in the number of tourists visiting the area, following the airing of an HBO mini-series entitled *Chernobyl*.

Recently, the Chernobyl area caught the attention of biologists, due to the revelation that animals are thriving in what humans consider a "dead" zone. In fact, the animals appear to be thriving better than when humans existed in the area in large numbers. Today, the deserted area is home to beavers, owls, moose, deer, lynx, wolves, brown bears, and tens of other animals. Despite the unsafe radiation levels, wildlife is increasing in number without human interference. Seemingly, radiation is friendlier to the animals than man.

Apart from wildlife, some of the domesticated animals left behind during the exodus gave birth to offspring which are still thriving in the zone. Dogs are the most dominant pets found in the exclusion zone. They appear to live normally. However, researchers have reported that, while radiation is likely to kill the dogs prematurely, they die of natural causes, such as harsh winters. This occurrence has shortened the lifespan of the dogs to just 6 years. On the brighter side, there are organizations that provide shelter, food, and medicine for the animals. Evidently, dogs are the only animals living within the deserted area that would flourish better if the humans returned.

Conclusion

Congratulations! You made it to the end of the book!

This book was accurately composed to provide you, the reader, with all possible details about the Chernobyl nuclear disaster. We started off by providing a short history of the Chernobyl area before the accident. You now have a clear understanding of where the town came from, and the purpose it was intended to serve.

Additionally, you now understand the nature of the RBMK-1000 reactor, which was at the center of the disaster. Probably, other books on this topic only talked of a reactor, leaving the readers trying to figure out how reactors work, and how this one led to the destruction of Chernobyl. We have provided detailed information and taken you to the center of the reactor, so you can easily explain how it was built, and how it is supposed to generate electricity. From there, the accident unfolds, and all the events that succeeded the event are discussed in the simplest way possible.

Hopefully, you found all the other information in this book to be, similarly, educational, engaging, informative and enjoyable. After reading the book, you probably realized that almost everything that a person needs to know about mankind's worst nuclear disaster is right here. Feel free to recommend it to anyone else, and, wherever possible, give it a good rating, so others can be as informed as you are now.

In conclusion, it is important to state that thorough research was conducted during the writing of this book. However, as the reader, you might come across information that appears doubtful or unclear. You reserve every right to correct any part where you feel a different source might have been included. All in all, you are promised that such occurrences are insignificant, if any.

Finally, please accept the utmost appreciation for purchasing and reading *History of Mankind's Greatest*

Disaster: A Walk Through The Chernobyl Nuclear Catastrophe. Hopefully, you enjoyed reading it.

Thank you very much!

it takes ONLY A MINUTE TO GIVE US A REVIEW

www.ingramcontent.com/pod-product-compliance
Lightning Source LLC
Chambersburg PA
CBHW030637220526
45463CB00004B/1561